MOLLI THE MAT

BEYOND MATH ANXIETY

99 Insights

(and a Calculation's Not One!)

Molli Osburn, M.A.

Copyright © 2018 by Molli Osburn, M.A.

All rights reserved. This book or any portion thereof may not be reproduced or used in any manner whatsoever without the express written permission of the publisher except for the use of brief quotations attributed to the author.

Published by Molli the Math Lady, L.L.C.
www.MollitheMathLady.com

ISBN: 978-0-692-14050-5 (Paperback)

Printed in the United States of America

Book Cover Design, Interior Design, Typesetting, and Pre-press Production: Lisa Von De Linde, LisaVdesigns.com

First Paperback Edition

DEDICATION

*This book is dedicated to all the adults
who resonate with the statement,
"I wish the Math Lady would have been around
when I was young!"*

*This includes my Aunt Lisa, who unfortunately
did not live to see the publication of this book.*

*While it might be too late for Aunt Lisa,
it is never too late for you.*

CONTENTS

INTRODUCTION

Welcome! You probably hate math! (Or know somebody who does.)

YOU PROBABLY PICKED UP this book reluctantly. You might have seen the word "math" in the title and your heart started beating faster, your hands started trembling, and you started getting knots in your stomach. Your brain might have been flooded with painful memories of math from yesteryear. You probably have a love-hate relationship with math, or to be perfectly honest, a hate-hate relationship with math.

If I had a dollar for every time an adult said to me, "I wish you would have been around when I was young!" I would have a fortune. Every time I hear that statement, my heart aches a little, knowing that so many adults today are scarred by their painful experiences of math education in childhood and adolescence. To make matters worse, many kids and teens today are also suffering in their math classes.

Perhaps you are the parent, grandparent, teacher, or counselor of such a child or teen, or perhaps you are such a preteen, teen, or college student yourself. On the other hand, you might have a child, grandchild, or student who seems to be a prodigy in math, and you can't seem to keep up with them. Either way, I've got you covered.

First and foremost, if you are an adult who has negative feelings about math, let me reiterate that you're not alone. The phenomenon of negative feelings about math is unfortunately all too common, especially among women. In our society, it has become as much a female bonding ritual to say that you're bad at math as it is to compare hair, clothes, and makeup. That is unfortunate, because math is an essential skill in our society, and is vital to many careers of the future. That is why most of these grown women do not want their daughters and granddaughters to suffer the way they did. So how do we break that cycle?

It is a complex question and one which we will explore throughout the 99 Insights. In the meantime, as a parent or a teacher, the first thing that you have to remember is that you are a role model for your children and/or students. In other words, you set the tone. If the tone you set is that of hating math, then they will likely follow in your footsteps. So, before you go to help your children, grandchildren, or students, it is vital that you explore your own relationship with math. Because really, what this book is about is exploring one's relationship with math. In spite of common beliefs, your relationship with math does not end when you graduate high school or college or when you become an adult, which is why it is worth it to explore it now (and I promise, no calculations or equations!).

Why we need a new approach

If you are an adult who is reading this book, then there is a good chance that you suffered in math classes during your childhood and adolescence. Conversely, if you are an adult who happens to be a math teacher, professor, scientist, or engineer who has always enjoyed math, then there is a good chance that you know a child or teenager who is currently suffering in math.

Let's talk about the common denominator here: suffering. So much pain, anguish, and anxiety accompany math today. Not just math, but the whole school system. Students today are under much more stress than students in the 1990s or early 2000s. There is more pressure, more competition, and higher standards, both in the school system and within themselves.

Many parents ask me about Common Core, and with good reason. It is the "buzzword" in education these days, especially in math and English. Most of the standardized tests are based on Common Core standards, so it is no wonder so many parents have concerns about Common Core.

One of the most common complaints that I hear about Common Core is that it is unfamiliar to the parents, and is different from the way they learned math when they were in school. In other words, Common Core is making parents suffer, just as it is making students suffer. My response to the questions about Common Core is that it is a legitimate concern, but it doesn't tell the whole story, because the theme of suffering in math goes well beyond Common Core. In other words, no matter what the current buzzword and methodology in education is, there will always be suffering in math, especially among girls and women. That is, until we break the cycle of suffering.

So, this begs the question. How do we increase achievement in math and science while decreasing suffering? And not only decreasing suffering but also decreasing sexism? This is especially unfortunate, because as former President Barack Obama and former presidential candidate Hillary Clinton have said on several occasions, STEM (science, technology, engineering, and math) careers are the wave of the future. And the more women we have in these careers, the better off we'll be. Unfortunately, with the great deal of suffering that current math education causes, we have fewer women in these fields, not more. This is not because girls have less aptitude in math or science, but because of societal stereotypes. These stereotypes hurt boys just as much as they hurt girls, by way of toxic masculinity.

Clearly, we need to do something. We need an approach that goes beyond the math material, and dives into the emotional and spiritual aspects of this suffering as well as the social and political climate in which math and science occur. In addition, we need to infuse more creativity into math and science, which is sorely lacking in the era of Common Core and standardized tests. We need an approach that considers the whole person, not just the math material.

The 7M Pyramid

In light of the aforementioned concerns, I developed the revolutionary 7M Pyramid, a system of looking at math in a holistic way that goes beyond the material, and looks at the whole person. It originally started as the 5M Pyramid, but after I had the privilege of taking Gabrielle Bernstein's Spirit Junkie masterclass in New York in 2017, I added the bottom two levels, which incorporate aspects of spirituality (not necessarily religion).

Before we go into the technicalities of the pyramid, I would like to add a brief note about spirituality and religion. I know that it is a divisive issue and that people have a variety of spiritual and religious beliefs, including atheism. If you already have a religious practice, feel free to use it as you work through this pyramid. Likewise, if you are an atheist or are skeptical about spiritual things, you can still use the pyramid, but feel free to omit the bottom two levels.

Now let's take a look at the pyramid.

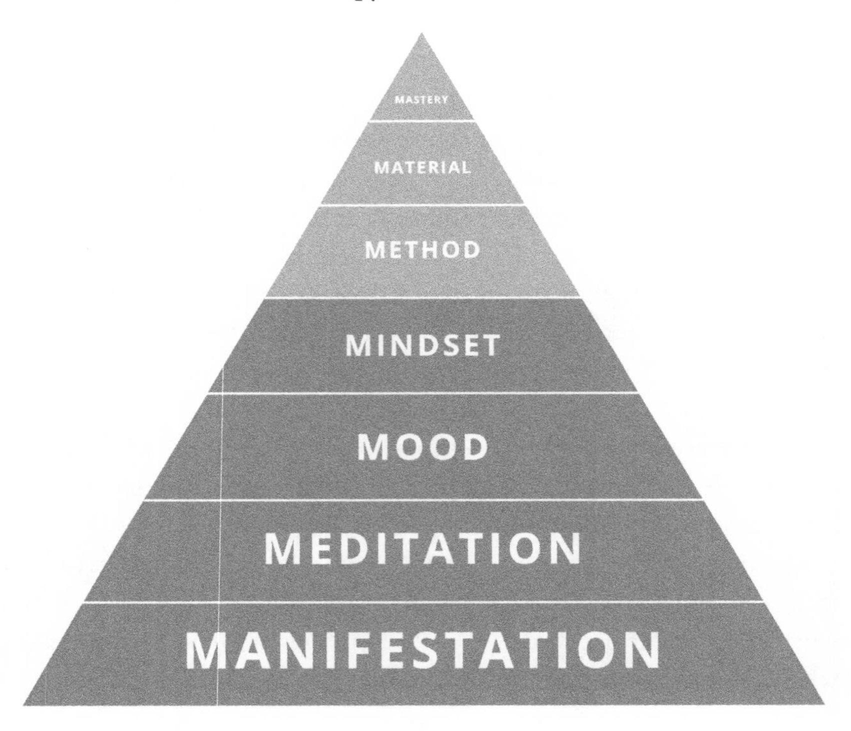

The top level is mastery. That is the result that you want, which in the case of math, is usually a grade or a test score. When parents come to me, they are almost always concerned with the results, which usually include grades and test scores, but might also include their child's level of understanding in math.

The second level is material, which is usually what people think of when they think of math. In other words, arithmetic, algebra, geometry, trigonometry, calculus, and other levels of math. Or within those levels, the particular content. For example, within arithmetic, it could include addition, subtraction, multiplication, division, or fractions.

The third level is method. Whereas the material focuses on what you're studying, the method focuses on how you're studying. There is often more than one method for every type of material. This level also includes study skills, organization, time management, and test-taking strategies. When parents ask me about Common Core, this is the level they're referring to— the material is basically the same as what they learned in childhood, but the method has changed, which is why they're frustrated and confused. This is a legitimate concern, but we still need to go deeper.

The first three levels (mastery, method, and material) can be grouped into what I call the strategy levels, which is what most tutors and teachers tend to focus on. The remaining levels are what I call the energy levels, and are vitally important as well.

The first energy level is mindset. This includes your thoughts about math. In other words, what are you thinking to yourself as you do the math? What is your self-talk like?

The next energy level is mood. This includes your feelings about math. Mindset and mood are inextricably linked because you can't have thoughts without feelings, and vice versa. In other words, your thoughts affect your feelings, and your feelings affect your thoughts.

After this, we have meditation. Here, we're moving into the more spiritual levels, so please feel free to skip this, even though I highly recommend it. In other words, meditation helps you to ground yourself, and to not let your negative thoughts and feelings affect you. We all have negative thoughts and feelings. Meditation is a tool to help you reframe those negative thoughts and feelings. In spite of popular beliefs, meditation is not just sitting and chanting. It can also include walking, yoga, journaling, listening to music, using essential oils, or any other activity that calms you down.

Lastly, we have manifestation, which is co-creation with the Universe. In short, we are all in vibration, and the Universe (or your higher power) picks up on our vibrations. In other words, if you're in a high vibrational state, you are more likely to achieve your goals. Unfortunately, most students, parents, and teachers are currently in a low vibrational state when it comes to math. That is why we need to shift the energy around math.

What makes my approach unique is that while most math tutors focus exclusively on the strategy levels (mastery, material, and method), I also incorporate the energy levels. Even if we skip the meditation and manifestation, the mindset and mood are vitally important, not only for students, but also for parents and teachers.

- Molli Osburn, M.A.
"Molli the Math Lady"

THE BASICS

(Insights #1-12)

Insight #1

Math anxiety is a very real problem, but it goes much deeper: math shame.

MOST PEOPLE ARE FAMILIAR with math anxiety. The sense of fear, dread, and trepidation that accompanies any attempt at solving a problem with numbers (or letters, in the case of algebra), especially if it is a timed test. The physical symptoms of sweaty palms, butterflies in your stomach, trembling hands, and your brain drawing a blank. If you are a current student, you might get these symptoms immediately before or during a test, or even just thinking about a test. If you are an adult who has been out of the classroom for years (or even decades!), you might still get these symptoms when you are calculating bills or taxes. These symptoms of anxiety usually have a deeper root cause.

Before we address the difference between math anxiety and math shame, let me start by saying that if you're looking for a workbook on how to solve algebra problems, work with fractions, attack word problems, or master Common Core strategies, you're probably going to be disappointed. Either that, or you're going to be secretly relieved. With the exception of a few insights towards the beginning, this book is not about the math material, per se. Rather, it is about dealing with your emotional response to the math material, whether you are a parent, teacher, or student. These insights can apply no matter what level of math you or your child or student is currently taking, or even if you've left the classroom years (or decades!) ago. After all, whether it's first grade addition or advanced calculus, the emotional themes are the same.

Most people think of math as a completely logical subject, devoid of any emotion. If they want subjects that are more emotional, they think of things like art, poetry, and psychology. In other words, the logical and the emotional are seen as a contradiction, and the common view is that emotions don't belong in math.

Despite its reputation as a non-emotional subject, math is a subject that is deeply emotional to many people. Indeed, it is those emotions that cause many people to avoid math, or to abandon it in high school or college, never to return, and carrying those emotions with them for decades after they leave the classroom. And the core of this emotional response is shame. In other words, feeling defective and flawed as a human being. Unfortunately, math often makes people of all ages (especially girls and women) feel defective. It is this shame that keeps many women away from STEM careers. While math anxiety is important to address, its root cause is almost always math shame.

Insight #2

The combination of math and spirituality is not a contradiction.

MOST PEOPLE WOULD PROBABLY THINK that math and spirituality don't go together. Of course, they might go together like ice cream and ketchup, or like sweatpants and high heels.

You might feel the same way about spirituality and math. In other words, there's a time and place for each, but not at the same time or in the same place. I'm here to challenge this assumption. One of my basic premises is that math shame is a spiritual wound that requires spiritual healing. This is not well-known, so most people rush to solve the problem with an academic "Band-Aid" (a.k.a. tutoring). This approach rarely works, and rarely addresses the underlying hurt and shame. After all, math is very logical. On the contrary, spirituality, at its very nature, is not logical. In fact, you might be a bit skeptical about this "woo-woo" stuff, and might even be a bit hesitant about a math professional who not only believes in it but also promotes it.

Let's face it. Life is a mess of contradictions. In order to fully appreciate the logical, we must embrace the seemingly illogical. Paradoxically, by embracing the irrational, we can have a better appreciation of the rational. In order to do well with the logical material, you need to be grounded spiritually. That might look different for different people, and that's OK. On the other hand, I can tell you what it does not look like.

Being spiritually grounded does not involve frantically studying, beating yourself up for not knowing the next step in an equation, punching buttons on the calculator in frustration, and making excuses to avoid studying. Yet, so many students, parents, and teachers skip the step of becoming spiritually grounded because they think it's not necessary, or that their time would be better spent on mastering the material. Unless you are spiritually grounded, the time you spend studying will not be as productive. In fact, as paradoxical as it may sound, becoming spiritually grounded can actually reduce the amount of study time necessary to achieve the result that you want.

Before I go any further, I should clarify: spirituality in this context is not necessarily the same thing as religion, but they can be complementary. One can exist without the other, and vice versa. If you have a religion that you follow (i.e., Christianity, Judaism, Islam, etc.), please feel free to incorporate this material into your spiritual practice. On the other hand, if you are not religious, it is not a prerequisite for incorporating these principles into your life and your transformation of math anxiety and math shame.

If you are an atheist, you can feel free to skip the sections on spirituality. But you might want to read it anyway and keep an open mind. Or you can feel free to read it out of order. Even if you're not comfortable with using the word "God," you might find a concept of a higher power that resonates with you. I bring this up because a lot of people who are involved in math and science fields are atheists. After all, the debate between religion and science is ancient. It has many striking parallels between the original concept I brought up, which is that of math and spirituality seeming to appear to be contradictory concepts, even though they are intricately connected.

In going back to your math anxiety, remember that when you try too hard to understand the logical, your illogical brain can take over. Sometimes you just have to honor the illogical and go with it.

Insights #3-5

#3 - *The AAS Triangle of math feelings consists of anxiety, avoidance, and shame.*

#4 - *Math shame leads to math anxiety, math anxiety leads to math avoidance, and math avoidance leads to more math shame.*

#5 - *The Math Lady's version of the "AAS Triangle" has nothing to do with sides, angles, sines, or cosines.*

IF YOU HAVE PAINFUL MEMORIES of the AAS Triangle from your days in geometry or trigonometry, you can breathe a sigh of relief. I am not going to be quizzing you on sides, angles, sines, or cosines. Remember how I said in the title that there would be no calculations? I'm living up to that promise. In fact, this AAS Triangle is far more important than anything you might learn in geometry or trigonometry or any other math class.

This is a new AAS Triangle, one that has to do with emotions surrounding math, and has nothing to do with math itself. It consists of three core emotions, which are linked to each other: anxiety, avoidance, and shame.

We have discussed math anxiety in previous sections. It is the feeling of hesitation and trepidation when faced with a math problem, and is not just cognitive and emotional, but is also physical. It can include physical symptoms such as sweating, shaking, rapid heart rate, and digestive disturbances. Anxiety keeps you in the "fight or flight" mode and keeps your limbic system overactive. This is especially unfortunate when you need your prefrontal cortex to be running the show. We all have experienced anxiety at times (not just about math), and we all know that it is an unpleasant experience.

Naturally, most people would want to avoid unpleasant experiences. This then leads to avoidance, which is exactly what it sounds like. That is, purposely avoiding something. This can look like procrastination, resistance, or classic excuses. In other words, you don't want to face something, so you keep putting it off. Examples of avoidance would include skipping a dentist appointment or waiting until April 14th to do your taxes.

Then our conscience often gets the better of us, whether we like it or not. That's where shame comes in. Brene Brown is an author and leading authority on the subject of shame, including the key difference between guilt and shame. Brene Brown defines guilt as doing something wrong, and shame as being something wrong. For example, if you get a speeding ticket, you might feel guilty for speeding, which you know is wrong. If you take it a step further and say that it means you're a bad driver and therefore a bad human being, then it is turning into shame.

We can apply the same principles of guilt and shame to math, in that many people of all ages, especially girls and women, feel math shame. In other words, feeling that they are flawed at their core for not living up to an arbitrary standard in math, and feeling that their math shame defines them. This principle can apply to shame about other areas as well, not the least of which is my parallel struggles with weight and relationships.

If you feel shame about something, then you might engage in behaviors that are the exact opposite of what common sense would tell you to do to solve the "problem" that the shame seems to be about.

For example, if you are overweight and have shame about your weight, you might overeat, even though common sense tells you to eat less and/

or make smarter food choices in order to lose weight. Similarly, if you have shame about money, you might overspend, even though common sense tells you to save and/or pay down debt in order to have long-term financial security. Lastly, if you have shame about a past relationship (or two or three or ten), you might avoid dating, even though common sense tells you to go on more dates to find a new relationship.

In applying this to math, many students who have math shame procrastinate in their math studies, even though common sense tells them to study more. Similarly, many adults who have math shame might have regrets about their past, and do nothing to move forward in their present, because they think it's too late. In both cases, we need to address the shame before we can effectively address the problem-solving. The first step to addressing this shame is becoming aware of it.

In addition to understanding shame, and how it can make us behave in paradoxical ways, we also need to understand how shame is related to anxiety and avoidance. Thus, the AAS Triangle was born.

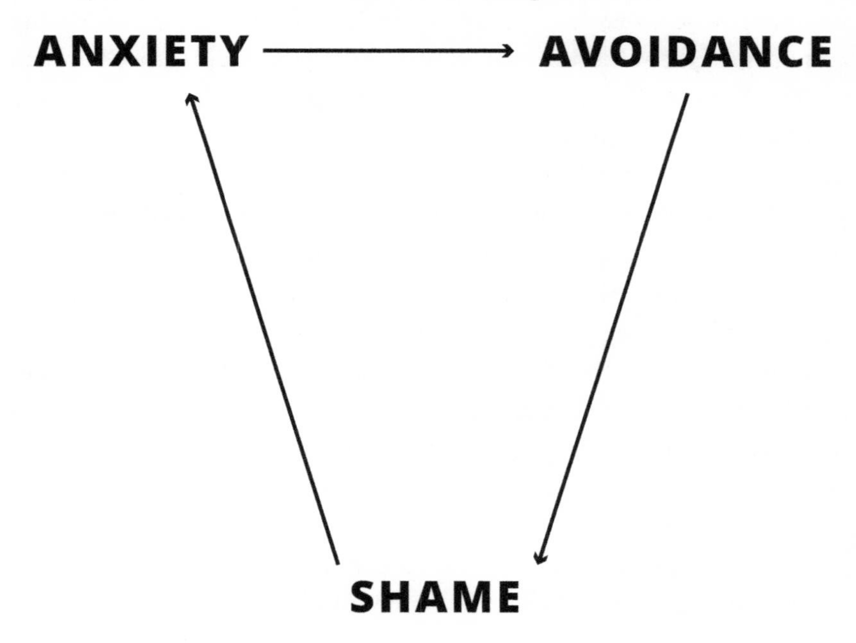

Here, anxiety leads to avoidance. Then, avoidance leads to shame. Then, shame leads to more anxiety, and the cycle repeats. Sometimes, the cycle repeats so rapidly, you don't know what point you're in. When it comes to the adults who lament, "I wish you would have been around when I was young!" they are most likely in a perpetual state of avoidance. In other words, they have experienced so much anxiety and shame surrounding math that they just want to escape the painful feelings, and avoid it all together. Hence, the reason why most of these adults have not taken math since high school or college, at which point they usually took the minimum level of math required to graduate.

If they attempt to get out of the avoidance state as adults, it triggers their underlying shame, which is why most adults who suffered in math as children are reluctant to revisit their feelings about math as adults. On the other hand, if their children, grandchildren, or students are having similar experiences in math, it might trigger those old feelings and might bring up memories that the parent or grandparent hasn't thought about in years. When that happens, the best thing to do is to be aware of those feelings.

In breaking the AAS Cycle, most tutors and teachers intervene at the avoidance stage, which usually involves telling the students to buckle down and study, do their homework, and stop making excuses. While this might work in the short term, in looking at the triangle, you will see that anxiety leads to avoidance. In other words, there is almost always an underlying emotional reason for avoidance. This usually has to do with anxiety. With this in mind, it would probably be more effective to intervene at the stage of anxiety.

Let's take it a step further. As we can see, anxiety is often triggered by shame. With that being said, it is most effective to begin by addressing the level of shame. Which isn't easy, but will yield the most effective results long-term. In other words, math shame is a spiritual wound that requires spiritual healing, and the shame is rarely healed by the common academic fix (i.e., tutoring). We need to get to the bottom of these limiting beliefs, fears, and emotional wounds, and reframe and heal them. That way, we cover all the bases of the AAS Triangle.

Speaking of the AAS triangle, you might have panicked when you saw the term at the beginning of this section. This just goes to show how deeply ingrained math anxiety and math shame triggers can be. If you were tempted to skip over this section because of your anxiety, then that would have been an example of avoidance. Like I said in the beginning, this has nothing to do with sides, angles, sines, or cosines.

That is a great example of using humor to alleviate math anxiety and math shame, and even math avoidance. If you don't recall AAS Triangles from geometry or trigonometry, that's OK, but in their most basic sense, they have to do with the relationships between angles and sides in a triangle. We also have another case, which is SSA. That is often an ambiguous case, which is why I tell my students that it can be a pain in what SSA spells backward! That, again, is another good use of humor in alleviating math anxiety.

All in all, the AAS Triangle is very important not just for math, but also for other areas in your life that you might be struggling in. A good example would be my past struggles with weight and body confidence, and with relationships and social confidence. The AAS Triangle fits perfectly with the 7M Pyramid, because it's not just about the material or even the method. Most of what's going on in the AAS Triangle has to do with mood. Of course, mindset and mood are inextricably linked, because you can't have thoughts without feelings, or vice versa. Spiritual tools can be an excellent resource for helping with anxiety and shame. For example, meditation can be great for calming anxiety. On the other hand, avoidance can make your mindset and mood worse, not better, in the long run, even though it might seem like it's making you feel better in the short run. So, we really need to start at the shame level. We need to heal those feelings of inadequacy and reclaim our power. Because this is what is missing in the current approach to math education. I work with you in all three areas of the triangle, and how they fit together.

Insight #6

The APR Method of letting go of math shame consists of awareness, processing, and release.

I'M SURE THAT to many of you, APR sounds a lot like AAS. For the adults reading this, you might think of the dreaded number on your credit card statement, or if you're under 18, you might think of a formula on one of your math tests. (For those of you who are incredibly curious, APR stands for annual percentage rate, but again, we won't be doing any calculations!). Like AAS, APR in this context has nothing to do with the math material. In this case, APR is a method that we can use to let go of math shame. The three steps are awareness, processing, and release.

As with anything, we need to start with awareness. Not only awareness of the existence of math shame, but also awareness of how it affects us. For example, noticing when something triggers us. Instead of trying to avoid triggers, become curious about them. Is there any pattern to them? Is there anything in particular that triggers them? For example, you might find a pattern of getting nervous the night before a math test. Fair enough. As they always say, awareness is half the battle. Not only becoming aware of your patterns or triggers, but also if you think you're anxious about one thing, you could be anxious about another. I remember one time when I was frustrated that my computer wouldn't work, and I was trying every little thing to get it to start up again, but then I realized that my anxiety was really about starting a new graduate program.

We can also become aware of physical sensations of math anxiety. How does it feel in your body when you take a math test? How does it feel in your body when you procrastinate? Again, there are no right or wrong answers here. The first step is just to become aware of your patterns and to nip them in the bud as they happen.

That brings us to the next step, which is processing. Once we are aware of these feelings and triggers, we can choose a new way. For example, what does this feeling mean? Where is it coming from? Does it have to do with a past experience? If it has to do with a past experience, perhaps some healing of that experience is in order. Or perhaps just seeing things differently will help.

Our last step is release. That is, to release the old feeling or belief that is holding you back. Sometimes what I do with my math empowerment retreats is to hold a "release ceremony." During the spring and summer, we often have a campfire during our retreats, and part of the assignment is to write the beliefs that no longer serve you on scraps of paper and then burn them. For example, the belief that you're bad at math, or the belief that girls aren't supposed to be high achievers in STEM, or the belief that women shouldn't earn more than their boyfriends or husbands (assuming, of course, that they're heterosexual). Even during the fall and winter, you can do this exercise with a fireplace (if you have one). Or if you can't safely burn them, then you can tear them up and flush them down the toilet (assuming your plumbing's up to code). Whatever you do, it is the symbolism that counts.

Remember, you are releasing what no longer serves you, and transforming it into a new, empowering belief. And that is the APR cycle (much more fun than the APR on your credit card!).

Insight #7

Racism and sexism cost opportunities in math and science.

BEFORE I DIVE INTO the topics of racism and sexism (and ableism and homophobia in later insights), I would like to start with a bit of a disclaimer. I realize that my audience has widely varying views on these topics, and on political topics in general. I realize that these topics are controversial, and can be a bit emotional. But for true healing to occur, whether in math or in any other area, we need to be willing to face topics that are uncomfortable and can trigger our emotions. That said, I would like to introduce this topic, which focuses on the theme of the importance of social and political issues in shifting the landscape surrounding math and science, and especially women in math and science.

There is no denying that sexism played a role in Hillary Clinton's loss to Donald Trump in the 2016 presidential election. There is no denying that racism played a role in the widespread dislike and obstruction of Barack Obama for eight years. There is no denying that there is still a lot of underlying racism and sexism in our political system and policies.

As they say, the political affects the personal, and vice versa. For example, the loss of talent in math and science may seem like a personal issue but can compound into a global issue. In other words, it becomes a political issue. Racism and sexism cost women and minorities a lot of opportunities in math and science.

The 2016 film *Hidden Figures* explores a lot of these themes. It was released right around the time of Donald Trump's election, which is perfect timing for highlighting the importance of being aware of racism and sexism. The film itself takes place in 1961 and is based on the true story of three African-American women who worked at NASA. Aside from math and science, another central theme that the movie highlighted was the racism and sexism that these women faced.

Unfortunately, racism and sexism are still rampant in 2018. Not only are they prevalent in society in general, but especially in the fields of math and science. In addition, racism and sexism can easily overlap, which is why African-American and Hispanic women are less likely than white women to go into math and science, and even less likely than white men. The same principle is true for gays and lesbians, and for men and women who have disabilities.

It should be reiterated that this is not because men are better or smarter at math than women, but rather because of the stereotype that they are. Ditto for stereotypes about racial and ethnic minorities, and stereotypes about gays and lesbians.

What do we do? Become aware of racism and sexism. Challenge it when you hear it. Stand up to the system. Most importantly, if you are a woman and/or a racial minority, become educated on these issues. And if you are a white man, own your privilege, and reach out and support the women and racial minorities in your lives.

Insight #8

In order to combat climate change, we need a new generation of STEM professionals.

TO REFRESH YOUR MEMORY, STEM stands for science, technology, engineering, and math. Needless to say, these fields go together intrinsically and cover a wide variety of topics. Just like with the sections on racism and politics, I must start with a disclaimer. That is, I am aware that people have varying opinions on global warming (a.k.a. climate change).

However, in this case, it is a matter of fact, not opinion. There is overwhelming objective scientific evidence that supports the assertion that global warming is a very real phenomenon. Not only very real, but also very dangerous for the future of our planet. Needless to say, we need to slow this trend down for the sake of our future generations. In order to combat climate change, we need many more science professionals. As former President Barack Obama and former presidential candidate Hillary Clinton have said on many occasions, STEM careers are the wave of the future. With so many talented young people (especially young women) dropping out of these fields, progress is a bit slower than it could be.

As most of you know, the Trump administration has undone many of the Obama-era regulations protecting the environment. In spite of overwhelming evidence that global temperatures are steadily rising, Trump and his team have shunned science in the name of protecting the financial interests of the coal and oil companies. In addition, Trump and his team have cut funding for scientific research.

In order to protect our environment, we not only need scientists and engineers but also citizens who are educated. The March for Science, which took place in April 2017, was a good start in educating the public about science. And websites like "I F***ing Love Science" are another good start. We need quality education in science for all people and not just STEM professionals. We also need to start at an early age, in order to get children (especially girls) interested in STEM careers.

Many of today's students are not well-educated in the workings of basic science. That is partially because science is frequently associated with math, and if a student has anxiety about math, then it is likely that they will be reluctant to study advanced science. Not only that, but math is also a prerequisite for many higher-level science classes, such as chemistry and physics. Thus, math anxiety can become science anxiety. Once again, we not only lose out on STEM professionals but also well-educated citizens in general.

Climate change is a critical issue for our future, which is why we need more STEM professionals to help solve it. Same principle with medical and technological advances. Our society, and our planet as a whole, cannot survive without science. Much progress in science stands to be lost, not only because of Trump-era policies but also because of math anxiety and math shame. In other words, while many people think of math anxiety and math shame as personal problems, they can compound into a social and political problem that affects us all.

Insight #9

Math shame can have large-scale effects on a social and political level.

AS WE SAW IN PREVIOUS insights, math shame leads to math anxiety, which leads to math avoidance. Thus, it is the last part that has the most profound effect on society. After all, when young women (and young men) drop out of college majors and career fields that involve STEM, it has a huge ripple effect on society.

This common hatred of math is a social and cultural thing. Even as young as seven, girls have been known to say that they don't like math. Indeed, by middle school, many of them have abandoned math and science in favor of fashion, makeup, dating, and social popularity. In other words, among heterosexual teenage girls, there is a misconception that boys don't like girls who are "smart," especially in math and science. Couple this with math anxiety and math shame, and we have millions of young women who avoid math and science before giving it a chance.

What are the societal effects of math avoidance? For one thing, we have fewer people in STEM careers. This leads to fewer advances in science and technology, which leads to less progress in global warming, health and medicine, and computers and technology. After all, we rely on medicine and technology every day. For the sake of our future generations, we need to act now to prevent global warming from progressing. That said, math avoidance can affect society on a global level in many ways.

The perpetuation of math anxiety and math shame in women reinforces the sexist stereotypes about women and math and science. These stereotypes are so ingrained in society that they even affect women in college and graduate school programs in math and science.

Some studies have been done, and have proclaimed that the age at which girls begin to lose interest in math and science is 15. Not 14, not 16, but 15. I beg to differ. Although they might proclaim that they are done with algebra at 15 (or 16), the seeds are planted a lot earlier. Math anxiety has been seen in girls as young as first grade, which is 6-7 years old. Less than half of the age suggested by the study. So, what gives?

It's a snowball effect. The seeds of math anxiety and math shame are planted at the beginning of formal schooling and continue to grow if unchecked. Thus, the study might have pinpointed 15 as the age at which math avoidance peaks. But we need to start much earlier and nip the anxiety, and especially the shame, in the bud. On the upside, it is never too late to overcome math anxiety or math shame, even if you have been avoiding math for years (or decades!).

Insight #10

There's no such thing as a "math person."

ONE OF MY BIGGEST pet peeves is when someone says, "I'm not a math person," or "I've never been a math person." It implies that math ability is something you either have or you don't and that you can't change it. In other words, it implies Carol Dweck's fixed mindset and not the growth mindset. For those of you who are not familiar with Carol Dweck and her work on mindset, the fixed mindset states that your abilities are set in stone and cannot be changed. Compare this to the growth mindset, which states that new skills can be developed over time. Thus, the common phrase "math person" implies the fixed mindset, in that a person either has math ability or they don't.

Let's rephrase this. Is there such a thing as a "reading person"? No, because reading is a vital skill to getting along in our society, and if anyone over the age of 8 (maybe 10 in some cultures) admitted that they didn't know how to read, people would judge them as illiterate. Nevertheless, it is common for adults (especially women) to proclaim that they're not a "math person."

This limiting belief also promotes sexist stereotypes, in that it implies that men are more likely to be "math people" than women. Similarly, in taking it a step further, many people believe that math ability is genetic and that you're either a "math person" or you're not because of your genes. Just as well, many mothers say that their daughters got their lack of math ability from them. Some of these mothers even take it a step further, and say things like, "your brother got his math ability from your father." What a way to disempower your daughter and to encourage sibling rivalry.

In an online course I was taking, I learned about the concept of "pedestals," in which we unconsciously believe that some people are superior to us, for whatever reason, and that we're lacking when we compare ourselves to them. Unfortunately, this line of thinking puts "math people" on a pedestal. When people meet me, they often say, "Wow, you're really smart!" or "I was never a math person!" These statements, or similar ones, put me and other so-called "math people" on a pedestal, which is not healthy thinking.

The thing is, math skills, like any skills, can be improved. That said, there is no such thing as a "math person." While math may come more easily or naturally to some people, that does not mean that those people are superior to those who might have to put in a little extra effort. In a Facebook meme that I recently saw about math mindsets, one of the suggested tweaks was changing "I'm not a math person" to "I'm not a math person— yet!" I suggested changing it to getting rid of the term "math person" altogether, and I got 50 likes on that comment. Because like we said, there is no such thing as a "math person," just like there is no such thing as a "reading person." Math and reading are both skills that can be improved over time.

Insight #11

There is a very real biological explanation for math anxiety.

WE'VE ALREADY ESTABLISHED that the old "study harder!" routine causes more anxiety and shame. There is a very real biological reason for this anxiety and shame. It has to do with the science of the human brain. Now, before you turn the page, please take out a pencil, put away your phone and laptop, and set a timer for five minutes.

Time for a pop quiz!

Just kidding! No quiz! (I bet you're relieved!)

Let me ask a simple question: How did you feel when I said there would be a pop quiz? Nervous? Frustrated? Down on yourself? Angry at me? What did you feel in your body? Racing heart? Sweaty palms? Knots in your stomach? It turns out, there is a very real physiological reason for your reactions, whether physical or emotional. And it all has to do with the brain. There are two basic parts that we want to pay attention to. The limbic system, which attaches to the spine, and the prefrontal cortex, which is up front.

The limbic system is very primitive and very quick. Think about our ancestors, living in caves. If a bear came after them in the middle of the night, would they have time to think? No, they would run for their lives. That is because the limbic system is designed to sense danger, and to flee from it!

This is where the fight or flight response comes from. From our limbic system. If any of you had sweaty palms or a racing heart or a knotted stomach when I said there would be a pop quiz, you have your limbic system to thank. The limbic system doesn't know the difference between a bear and a math quiz. The limbic system is not very smart, but it is very powerful. It screams at you. DANGER! FOOD! SLEEP! Compare that to the prefrontal cortex, which is responsible for thinking and analyzing. You need your prefrontal cortex to solve a math problem, or to read Shakespeare, or to remember what year Lincoln was shot, or to know whether or not something truly is dangerous. In short, your prefrontal cortex is very important, especially for critical thinking.

The catch is, while the limbic system SHOUTS, the prefrontal cortex whispers. That is why, when you're having anxiety about a test, it is pretty much useless to try to reason your way out of your anxiety. It's like your prefrontal cortex is whispering, "it's just a test," "it's ok if you don't know the answers," "your life's not going to be over if you don't get a perfect score." In the meantime, your limbic system is shouting, "YOU ARE A FAILURE!" "YOU CAN NEVER DO MATH!" "YOU'RE GOING TO WORK AT STARBUCKS FOR THE REST OF YOUR LIFE!"

No matter how irrational your limbic system may be, it's going to win. That is why, in order to successfully combat test anxiety, we must work with the irrational, not against it. Math anxiety and math shame are rooted in the limbic system, and no matter how hard we try, we cannot use the prefrontal cortex to reason our way out of it. In fact, the harder we try, the more our limbic system rebels. So that is why the key is to work with the limbic system, not against it. In other words, you cannot reason your way out of math anxiety or math shame.

Insight #12

Anxiety is a normal, evolutionary response to stressful situations. This includes math.

REMEMBER THE BIOLOGY of the human brain, and the limbic system and the prefrontal cortex? Our brains were evolved that way, to serve our ancestors. If an ancient caveman or cavewoman saw a bear, would they have time to analyze the situation? Probably not, because they would be in immediate danger. That is why the limbic system is so fast and automatic. It is simply because it is designed to keep us from danger.

Unfortunately, the limbic system does not know the difference between a bear and a math test. All the limbic system knows how to do is to protect us from danger. Whenever the limbic system senses a threat, it causes us to react. What served our ancestors might not serve us today. Same principle with wisdom teeth and appendixes. Who has had their wisdom teeth out or has had appendicitis? Those are examples of things that might have served our ancestors but are not useful today. And not only not useful, but also downright painful (or just a minor inconvenience).

It is the same principle with the limbic system. While it definitely served a purpose in the ancient days and still serves a purpose today, that purpose can sometimes get annoying and painful. Not just annoying and painful, but also counterproductive, as is the case with math anxiety. That is unfortunate because the limbic system can come with some pretty nasty side effects, like being much louder than our prefrontal cortex.

Imagine that you and your friend are in a crowded stadium for a concert for your favorite band or musician. You try to whisper something to your friend. Does he/she hear you? Probably not, because there's all this shouting in the stadium, and the band playing. Same principle with your prefrontal cortex and your limbic system. The limbic system is so much louder so that it drowns out the prefrontal cortex. That is why it is useless to reason your way out of math anxiety and math shame. Thus, we need to calm the limbic system before we can actively work with the prefrontal cortex. In other words, the brain science between the energy and strategy principles. Thus, the energy work targets the limbic system, and the strategy work targets the prefrontal cortex. Too much strategy work and too little energy work can backfire and can make the energy worse.

That's why it is especially important to remember these evolutionary principles while doing our energy work. That, and sometimes, the limbic system is overactive in some people. That is where we have to draw the line between normal math anxiety (or any other anxiety) and an anxiety disorder. Once again, I am not a psychologist or psychiatrist, so I can't make that call. I will be discussing anxiety disorders at length in my next book, releasing in 2019. If in doubt, please see a psychologist or psychiatrist.

BEYOND THE BASICS

(Insights #13-22)

Insight #13

Math anxiety is a symptom, not a disease.

A LOT OF TIMES when people hear what I do, they want me to "fix" or "cure" their child's math anxiety. I wish it were that simple, but that's not how it works. It's not like a TV. You can't just rearrange a few wires, flip a switch, and then it's magically fixed. It's human nature. There are a lot more subtleties.

I like to say that working with math anxiety (and math shame) is more about healing and releasing than about curing and fixing. In addition, the healing process is not linear. It goes in all different directions. Like I said before, math anxiety is a symptom, not a disease. In other words, there is almost always a root cause.

Earlier, we talked about the connection between math anxiety and math shame, and the AAS Triangle of anxiety, avoidance, and shame. While they all deal with the "mood" level of the pyramid, shame bleeds a lot more into the spiritual levels than does anxiety. While I am not discounting anxiety at all, I am showing how math shame goes deeper than math anxiety and wounds us at a spiritual level.

Math shame is a spiritual wound that requires spiritual healing. Unfortunately, most parents and teachers are not aware of this, and so they think that the solution is a quick academic "Band-Aid" fix (a.k.a. tutoring). Ironically (and sadly), this quick fix often makes things worse, not better. Remember the 2015 Taylor Swift song "Bad Blood"? There's a line that stands out: "Band-Aids don't fix bullet holes."

While slightly exaggerated, its point is clear. In applying it to math shame and math anxiety, we can see the math shame as the wounds that are inflicted in childhood. Obviously, a "Band-Aid' fix is not going to heal them overnight.

This is why my work is so significantly different from traditional tutoring. While I don't discount the math material, it is far more important to focus on the lower levels of the pyramid. Although the bottom two levels (meditation and manifestation) are optional for those who are skeptical about the spiritual aspects of my work, the mindset and mood levels are essential. This is because your thoughts and feelings about math directly affect your capacity to learn the material in math. That, in turn, affects your performance in math.

The material is still vitally important. After all, the math material is what's going to be on the test, and is what makes up math itself. But in order to get results, you sometimes need to look deeper, at what is causing the symptom. Something is obviously blocking you from getting the results that you want, and in order to get the best results, it is worth the time and effort to discover what that block is, and to work through it. Otherwise, if these blocks are ignored, they only get bigger, and your desired results become more and more elusive.

When math anxiety creeps up, instead of trying to get rid of it, be curious about it. Invite it in. What is it trying to tell you?

Insight #14

Your success in math is approximately 30% strategy and approximately 70% energy.

THIS IS WHERE I DIFFER significantly from most math tutors and supplemental education centers. Not only that, but I also believe that this part is missing in the current educational system, as well as the educational system of yesteryear. In going back to the 7M pyramid, most education focuses primarily, if not exclusively, on the levels of material and method, with a high emphasis on the level of mastery (a.k.a. grades and test scores). In other words, there is a primary focus on the levels of strategy. The levels of energy are rarely, if ever addressed, and if they are, they are considered a supplement, or a luxury.

This is unfortunate, because most people, including parents, teachers, and students, are not aware of the importance of one's energy in setting the tone for learning math. Energy is vitally important for both students and parents, and students' and parents' energy can affect each other.

You might be thinking that this is all a bunch of "woo-woo" stuff, but there have been scientific studies proving the existence of vibrational energy. In going back to a purely technical standpoint, all matter in the universe is in vibrational motion. This includes living things, such as human beings. We are all in vibrational motion, and at any given moment, we are all in a certain vibrational state. This vibrational energy can be given off, and others around us can pick up on this vibrational energy. If one is in a low vibrational state when it comes to math, then they will not be working up to their full potential.

That is why it is vitally important to become aware of your energy and to protect your energy. If you find yourself in a negative energetic state, or if you find yourself picking up on your parents' or teachers' negative energy, what do you do? When I find myself in a low vibrational state, I find it helpful to take a break from whatever I'm doing, and collect my thoughts. Then I breathe. That is vitally important, especially in the middle of a test when you're having anxiety. Just as well, if you're a parent and you're freaking out, it might help to take a moment to pause before you interact with your child or teen about the subject that is triggering you.

Once again, energetic boundaries are vitally important. I know that in our society, there is a toxic energy surrounding math, and part of my mission is to bring awareness to that toxic energy and shed a light on how to shift it. In order to shift it, we need to be aware of it. That is why the energy levels of my pyramid are so important, and why if you don't shift the energy levels, then the strategy levels will not be as effective.

P.S. - The 70/30 ratio that I came up with is arbitrary, and varies with each individual and each situation. In most cases, though, it is significantly different from 50/50, with energy carrying more weight than strategy.

Insight #15

In order to successfully learn math, you must first un-learn your negative thoughts and feelings about math.

IT IS COMMON KNOWLEDGE that most people come to me to learn math. In other words, to learn the material. In light of my emphasis on shifting energy around math, an important part of my process is "un-learning" your negative thoughts and feelings about math that are no longer serving you. That requires a lot of emotional and spiritual processing. In this case, it is vitally important to have someone to hold space for you. By holding space, I mean not judging, fixing, or offering unsolicited advice. Rather, just being there as a sounding board, and offering encouragement and support.

Unfortunately, we learn a lot of toxic, limiting beliefs about math from our culture, and are bombarded by those beliefs every single day. Whether or not you are still in school, you are exposed to these negative beliefs, and they can be unconsciously internalized.

In going back to our pyramid, you can also think of it as building (or remodeling) a house. The energy is the foundation, and the strategy is the decorations. In order to get the best results, you need to start with the foundation, no matter how tempted you might be to start with the decorations. In other words, no matter what material you're learning (or what decorations you want for your house), it's not going to work unless you have a solid foundation. Think of it as being a "deep-clean" for your beliefs about your math abilities. Lots of these beliefs are passed through the generations, so in a way, you're becoming a pioneer, or a trailblazer, for the next generation.

The real work comes in challenging the old beliefs. A lot of these beliefs are unconscious. And we can't pile new math material on top of old faulty beliefs. This is where the real work comes in. We need to dig deep and dismantle these old beliefs. Think of it as a scavenger hunt of sorts. We need to deconstruct the past, and then reframe it.

The psychological principles of learning can apply just as easily to "unlearning" old beliefs as they do to learning new material. One such principle is that of classical conditioning. Ever heard of Pavlov's dogs? Pavlov was a psychologist at the turn of the 20th century, and he did an experiment with dogs, to see if they could learn to salivate without food. Long story short, the dogs learned to salivate when a bell was paired with their food, and they eventually salivated when the bell was rung without the food. Thus, they learned to automatically associate the bell with food. A similar principle occurs with anxiety and math tests. If self-doubt is paired with a math test, then anxiety will be an automatic response whenever there is a math test. We need to unlearn that response.

So how do we do that? First, become aware of it. Then, try to associate doing math with something pleasant, like a pleasant sound or a pleasant fragrance. This will then de-activate the limbic system and will help you focus on the task at hand.

Insight #16

Passion creates a healthy mindset and mood. Pressure creates an unhealthy mindset and mood.

IN ADDITION TO THE 7M'S, I also have several P's, half of which support a healthy mindset and mood, and half of which are detrimental to a healthy mindset and mood. The P's that foster a healthy mindset and mood are passion, purpose, and perseverance. The P's that foster an unhealthy mindset and mood are pressure, perfectionism, and procrastination. We will address perfectionism and procrastination in more detail later, but for now, know that they're both generally unhealthy for your energy. That is because they're both related to pressure, which is the opposite of passion. Most importantly, let's talk about how we can reduce pressure and replace it with passion.

One of my biggest pet peeves is a phrase that while well-meaning, implies pressure. It is commonly said to students by parents and teachers who think they just need a little extra push. This phrase is "Come on!" This phrase bothers me because it implies pressure, rushing, and cajoling. It puts students on the spot and activates their limbic system. It invalidates whatever emotions the student is feeling at the time. These emotions might include but are not limited to, fear, shame, anxiety, and reluctance. In other words, "come on!" feels icky, and creates a sense of pressure.

What can we say or do instead of "come on!"? One of my favorite tools when coaching a person (whether or not a student) who seems reluctant to do a task is to ask, "would you be willing to (fill in the blank)?" This puts the power back in their hands. Of course, in the school system, it might seem like they don't have a choice, but if you think about it, there's always a choice. We always have choices, but those choices have consequences. If your student makes the choice to say “no” to the assignment or test at hand, that’s their choice, but they’re responsible for whatever the consequences may be (i.e., getting a lower grade, failing the class, or even not graduating).

Sometimes a “no” is really a “not now.” If that’s the case, ask when they would be willing to have you check in again. After all, sometimes a person is having an “off” day or even an “off” moment. If that’s the case, it might just take a quick fix to get them back into alignment. But don’t force it.

Instead of pressuring a student to do their math work, let’s connect it to their passion. While it might be a challenge to cultivate a passion for math in and of itself, it might be easier to connect math to a passion that a person already has. For example, I once had a student who was studying to be a veterinarian, but she was struggling with the calculus requirements. By connecting the calculus requirement to her passion for animals, we were able to get her through a subject that she might have otherwise found unpleasant. All in all, finding passion is about finding joy. By getting yourself in the high vibrational state, you can much more easily attempt the math that might otherwise seem challenging.

Insight #17

There is a major difference between compassion and sympathy.

COMPASSION IS VITAL in healing math anxiety and math shame, and parents and teachers play a key role when dealing with students who are suffering from these challenges. Notice that I didn't use the word "sympathy" in place of compassion. That is because there is a world of difference between sympathy and compassion.

Remember that helping someone (including yourself) with math anxiety and math shame is an exchange of energy. In other words, it is an exchange of energetic states, and of vibrational states. Remember what we said about low vibrational states versus high vibrational states? Well, sympathy literally means taking on the feelings of another. In other words, sympathy means that when the person you are trying to comfort is in a lower vibrational state than you, then you are bringing yourself down to their vibrational state. In other words, you are taking on their feeling, and are making yourself feel worse in the process. By being sympathetic, although well-meaning, you are in a sense, making yourself less useful, and are making yourself suffer in the process.

By contrast, compassion means attempting to raise the person you're trying to help to your own higher vibrational state. Or at the very least, you are staying in your own higher vibrational state while they wallow in their lower vibrational state until they are ready to raise their energy. In other words, compassion means showing caring and concern for the other person but protecting your own energy. Compassion involves energetic boundaries.

Unfortunately, many parents and teachers who haven't healed their own math shame from years ago might inadvertently go into sympathy mode with their children or students. Although they are completely well-intentioned, and genuinely want to help their children and students, they are making themselves suffer in the process. For example, if a teenager comes home and tells her parent about a horrific experience with a math test, the parent might have a flashback to a similar experience in their own adolescence. Then they tell their teenager the story and relive the feelings that they had 20 or 30 years ago. Thus, the parent is joining the teenager in the lower vibrational state.

Once again, if you have done something like that, you are only human, so please be gentle with yourself. I am using this as an invitation for teachers and parents to be aware of the difference between sympathy and compassion and to practice compassion. In other words, listen to your children and students, validate their emotions and concerns, but ultimately, protect your own energy.

Ultimately, protecting one's energy has to do with self-care and boundaries. In other words, have an outlet of sorts, and a way to recharge. That could look like journaling, meditation, exercise, or creative pursuits, to name a few. In other words, don't wallow in the negative emotions. Remember, your students and children need you to be at your best energetic self!

Insight #18

There is nothing wrong or shameful about asking for help or support. This includes help and support in math.

MANY TIMES, STUDENTS (and sometimes parents) see a stigma in getting a math tutor, like somehow, they "failed." As we discussed before, failure is a part of the learning process, and there's no such thing as a person being a "failure." But in our culture of comparison, it is easy to feel "behind" and lacking. Similarly, many parents might feel that their child needing a math tutor is a poor reflection of them as a parent, and that might trigger their own feelings of shame. There is a similar stigma to going to see a therapist, as well as taking medications for a mental illness. Taking medications for a mental illness is no different from taking medications for a physical illness, and there is little to no stigma for that.

Just like we need to reduce the stigma associated with mental health care in our society, we also need to reduce the stigma associated with getting support for academic subjects, including math. The traditional tutoring paradigm often adds to that stigma, with its emphasis on the material, and rarely addressing the limiting thoughts and feelings. Thus, as students get more and more frustrated with the material, their limiting beliefs and fears amplify and are rarely addressed. By taking a bottom-up approach through the 7M Pyramid, we are able to transform these fears and limiting beliefs.

Unfortunately, these principles are not well-known in society, or even in the educational system. If anything, our educational system perpetuates this stigma, and makes the fears and limiting beliefs worse. Thus, our educational system is a trigger for a lot of toxic energy. It is my goal to spread the word. Not only to de-stigmatize getting help and support for math anxiety and math shame, but also to change how society sees these phenomena.

Because it is all interconnected, and there are many layers. Once again, it's not just about understanding the material or getting a good grade. It's not just about the children, either. The first layer is grades and test scores. Then we go deeper, into understanding the material. Then we have math anxiety and math shame in students, followed by math shame in adults (including parents and teachers), and lastly, math shame in society (which includes sexist and racist stereotypes about math ability). As you can see, it is all interconnected, from the personal to the social. It is at the societal levels that these stigmas get started. The first step to breaking the stigma is to simply become aware of the stigma, and how it affects everyone.

Insight #19

In order to empower the next generation of STEM professionals, we need to bring compassion and joy back to math education.

MOST PEOPLE DO NOT FIND math to be enjoyable. One of my mentors was doing an online seminar on "finding your joy," and when I said that I found joy in math, she and several of the other women in the seminar gave puzzled comments and emoji's. That is not surprising, because, for most people, math is a source of pain and dread. This pain and dread goes back to childhood, and continues into adulthood, and often lasts a lifetime.

There is a lot to be learned from my mentor and her seminar. In addition to focusing on joy, she also focuses on compassion, which goes hand in hand with joy. So, what is joy? What is compassion? Joy is our natural state and is a state of being one with oneself, and one with the Universe. It is a high vibrational state and a state of natural creation. People might find joy in various things, such as spending time with loved ones, creating artwork, reading a book, taking a walk in nature, or playing a musical instrument. Similarly, compassion relates to joy, because it is also a high vibrational state in which we are deeply connected to others on a high level.

How do we apply those concepts to math education? Remember, empowerment comes from a place of joy, and also a place of compassion. In order to achieve true empowerment, whether for men or women or for any age group, we need to be in that high vibrational state of connection to others. We also need to add in the concept of forgiveness, which is key to compassion. And not only the forgiveness of others, but also forgiveness of oneself. Because remember, anxiety and shame are low vibrational states. In order to increase achievement and empowerment, we need to be in a high vibrational state.

Unfortunately, with the way our educational system is structured, that is easier said than done. So how can we make STEM joyful? The introduction of STEM toys and field trips for young children is a good start. Introduce them to the wonder of the natural world, and the joy of curiosity and forming patterns. These might work for young children but are often marred by math anxiety and math shame starting in the middle school years.

As Pablo Picasso once said, "the creative adult is the child that survived." The key is to nurture the inner child and not erase that sense of wonder. Even as tests and pressures pile up, it is vital to retain that child-like sense of joy and wonder. Not only that but also to have compassion for oneself and others. Because, ultimately, STEM careers are an act of service to the next generation, which is an act of compassion in and of itself.

Insight #20

Meditation can go a long way in releasing anxiety and shame before a math test.

IN GOING BACK TO THE 7M Pyramid, you'll notice that meditation is the sixth M. In other words, it's towards the bottom of the pyramid, in the spiritual realm (beyond mental and emotional). Just as well, in terms of the head, heart, and knowing, meditation relies on your knowing, or your intuition. Nevertheless, meditation can also support your head and your heart. In going back to the pyramid, meditation can be an excellent tool to support a healthy mindset and mood, as well as greater understanding of the methods and material, and ultimately, more mastery (aka better results).

I know that some people might have some resistance to meditation. Let me assure you that when I suggest meditation, I am not suggesting anything religious. Of course, if you have a religion that you already practice (Christianity, Judaism, Islam, etc.), please feel free to incorporate the practices of your own religion into your meditation. Even if you are an atheist, you can still benefit from meditation. That is because meditation not only connects us to our higher power of understanding (if any), but also connects us to ourselves.

Meditation is clearing one's thoughts, and getting out of one's head. Although math is traditionally thought of as a "head" subject, overthinking can lead to anxiety and self-doubt. Meditation puts you in the present moment, away from the pain of the past or the worries of the future. Meditation is not just sitting and chanting. Walking can be meditative, and so can listening to music, or coloring, or even cleaning. And journaling after meditation can be especially powerful.

In Julia Cameron's book *The Artist's Way*, one of her basic tools is morning pages or 2-3 pages of handwritten (not typed) journaling every morning first thing. This sets the tone for the day and helps to clear your thoughts. In the case of math anxiety, this can be helpful in reframing your thoughts and clarifying your feelings. Even so, it gives you a structure and a routine to your morning. Because your morning routine is key. For example, my friend, who's a life coach, once had a client who was trying to quit smoking. For her client, the first cigarette in the morning always tripped her up, so my friend (the coach) suggested that she create a morning routine that didn't involve cigarettes. The woman then took a kickboxing class and did her kickboxing routine in the morning instead of lighting a cigarette. Similarly, I used to check my phone first thing in the morning. Since replacing checking social media and email with morning pages and meditation, I have noticed a major shift in how I approach my day.

How about you? What can you add to your routine to release your anxiety and shame? More importantly, what can you subtract from your routine? Some unproductive behaviors might or might not be contributing to your anxiety, whether or not you're aware of it. Meditation can be an excellent substitute for some of these behaviors.

Insight #21

Each of the seven chakras in the body has a connection to math anxiety and math shame.

ONCE AGAIN, THIS IS A BIT "woo-woo," so please feel free to skip it over. As always, I am a firm believer of taking a holistic approach to moving through math anxiety and math shame. The fact is that these fears and shameful memories live in the body, not just in the brain. That is why we need to take a whole-body approach.

Much has been written about the seven chakras, or energy centers, of the body for centuries. As far as I know, this is the first iteration of how they specifically apply to fear and shame surrounding math. Here's an overview of each of the seven chakras, as well as how each one can affect math anxiety and math shame.

The Root Chakra represents safety and being grounded. It is often represented by the color red. This chakra has to do with math anxiety because when one is in the throes of worrying about math, that person usually does not feel safe and grounded. In other words, they see math as an immediate threat.

Moving up the body, the Sacral Chakra is represented by orange. It represents creativity, which is vital in problem-solving, not just in math, but in life in general.

Next is the Solar Plexus Chakra, which represents personal power and drive, and is located in the abdomen. This is the most intellectual of the chakras, and so it might be assumed that this chakra is the most associated with math. In order to take a balanced approach, we need to include all chakras. Disturbances in the Solar Plexus Chakra can often manifest as digestive upsets. As we all know, stomach problems are often a symptom of anxiety (not just math anxiety).

Next, we have the Heart Chakra, which represents love, compassion, and emotion, and is located in the center of the chest. Now once again, some might think that math and emotions are a contradiction, but if you have read all of the preceding insights, you know that there is a great deal of emotional complexity involved in math anxiety and math shame. Compassion not only for others but also for oneself, can go a long way in healing math anxiety and math shame. The Heart Chakra is vitally important for our work with emotions around math.

The Throat Chakra, which is located in the neck, deals with communication, both written and verbal. Remember when your math teachers always told you to "show your work"? Well, that's basically communication in math. Not only in the math material, but also in the thoughts and feelings surrounding math, communication is vital.

Next, we have the Third Eye Chakra, which is located between the eyebrows and deals with intuition. Remember when we talked about head, heart, and knowing? Well, this is the knowing part. This is the little voice that tells you what to do, even if it doesn't seem logical. This relates to math anxiety and math shame because a lot of intuition is involved in working through your emotions about math. Also, you know more than you give yourself credit for!

Lastly, we have the Crown Chakra, located at the top of the head, which deals with spirituality and our connection to a higher power. Here, we can rely on a higher power to help carry us through math anxiety and math shame and help us manifest our deepest dreams and desires.

All in all, the chakras are meant to be seen as a system, and not separately. In breaking it down into each individual chakra, we can identify where in the body blocks are located, so we can release them.

Insight #22

It is best not to study in the 24 hours leading up to a major test (SAT, GRE, etc.), if possible.

I KNOW IT SEEMS LIKE a blasphemy for a math tutor and test preparation coach to recommend that students not study. In reading the fine print, though, this only applies to the 24 hours before a major test. I am probably still going to be very unpopular among students, and especially parents. After all, conventional wisdom states that the more studying the better, and that the material should be fresh in your brain before a major test.

As paradoxical as it may seem, it is best to give your brain a rest in the 24 hours before a test, especially a major test such as the SAT or GRE. Remember the findings on the brain science behind math anxiety? Turns out, last-minute studying can activate the limbic system, and not in a good way. In other words, cramming only makes you more anxious, and makes you less likely to do well on the test.

Instead, you should relax and do something fun for those 24 hours. Although you should definitely study (preferably daily) in the weeks leading up to the test, the day before is a time to reset your brain. Do something that's fun for you, and take care of yourself. Play a sport, walk your dog, watch a funny movie, have an at-home spa day, do a creative project, or just act goofy. Don't be afraid to act like a kid!

Because of this reason, I do not offer strategy sessions in the 24 hours leading up to a test. Remember the pyramid? Studying in the 24 hours before a test promotes an unhealthy mindset and mood. This is especially true if studying replaces self-care during those 24 hours before the test.

What is self-care? It is all about achieving a balance. Unfortunately, in today's test-driven culture, there is a lack of balance. In other words, skewing the balance towards cramming for the test material, and not balancing it with the mindset and mood. But like we have said before, the mindset and mood are just as important, if not more important.

On my website, MollitheMathLady.com, I have several tips for the days leading up to a major test, with a special focus on the 24 hours leading up to the test. The most important thing to remember is not to study in the 24 hours leading up to a test. I know it's tempting, so I would advise that you put your books and notes away, or disable the app on your smartphone or laptop if you have high-tech study guides.

Then, get a good rest. I cannot overstate the importance of sleep. In addition to self-care and physical activity, sleep is vital the night before a big test. Ditto for nutrition. Often, when students are cramming, healthy eating and exercise habits can go out the window. But these are vitally important for a healthy mindset and mood. Once your body is well-rested and well-fueled, then you'll have the right energy to take on your test!

THE IMPORTANCE OF MATH

(Insights 23-31)

Insight #23

Addition, subtraction, multiplication, and division are arithmetic. Math is much more than arithmetic.

I CRINGE EVERY TIME someone says that they're "bad at math." Like we said in previous sections, this is a mindset, and a limiting belief that costs many women STEM careers. More importantly, even if you don't plan to go into a STEM career, some basic number skills are vital for your everyday life.

A common misconception is that arithmetic is the same thing as mathematics. Nothing could be further from the truth. By its strictest definition, arithmetic is about numbers and the manipulation of quantities. This can include addition, subtraction, multiplication, and division, and can also include fractions and decimals. Mathematics in its strictest sense is all about theory and patterns. In other words, it is more abstract.

This can cause a problem at approximately the middle school level, at which students typically transition from arithmetic to algebra. In other words, it becomes more abstract and involves more critical thinking. At this juncture, it can seem overwhelming to transition from numbers to letters. In fact, there are many jokes about the devil being the one responsible for infusing the alphabet into math, which basically represents the transition from arithmetic to algebra.

Abstract thinking abilities usually develop around age 11-13, and occasionally earlier in gifted students. That said, the transition to algebra is usually developmentally appropriate. There are some whose development may vary, and for these students, the transition to algebra can be particularly challenging. This is especially true for students who may have learning challenges. In this case, it is vitally important to provide support for the transition to more abstract thinking.

If you're like many people, you may have skipped ahead to this section. After all, this is the first section of the book that contains any reference to the material that is commonly taught in schools. In going back to the pyramid, we can see that the material is only a small amount. Learning the material is vitally important to achieving the results (mastery) that you desire. Therefore, an understanding of how the material typically progresses is vital. That said, it is vital to understand the difference between arithmetic and true mathematics.

Another mathematical transition typically occurs in grades 4-6, and involves the transition from whole numbers to fractions and decimals. In other words, the transition from whole to part. In this case, visual representations can also be helpful. For those who may have a unique developmental timetable, emotional support is critical.

Just like with life transitions, math transitions can be emotionally challenging. They can bring up old stuff, and old limiting beliefs, and can trigger fears and insecurities. Another part of it is fear of the unknown. Therefore, an awareness of transitions can go a long way.

Insight #24

Arithmetic is the foundation for algebra, geometry, trigonometry, calculus, and more.

IN BUILDING UPON the last section, arithmetic is vitally important for higher levels of math. Without a solid command of arithmetic, there is very little chance of succeeding in algebra, geometry, trigonometry, or calculus. Even statistics, which is more relevant to social applications, still requires a basic knowledge of arithmetic.

For these reasons and more, it is vitally important to develop a solid command of arithmetic skills, ideally when one is in elementary school. However, if these skills are not mastered in elementary school, it is not too late to master them during adolescence, or even adulthood. Just like literacy is vitally important to succeeding in our society, so is numeracy.

In many circles, the word "memorization" gets a bad reputation, especially when it comes to arithmetic. We must also consider that there is something to be said for having a familiarity with basic facts such as simple addition and subtraction, as well as times tables. The same can be said for basic fractions, decimals, and percents. For example, knowing that 1/2 is 50 percent, 1/4 is 25 percent, 1/10 is 10 percent, etc. Just a basic understanding of these concepts can go a long way.

There is much debate about how to teach these concepts. In general, I usually say that whatever way works for each individual student is usually the best way. In going back to the 7M Pyramid, this covers the level of "method," or the strategy or study skills. In other words, when learning arithmetic, it doesn't matter how you learn it, as long as you learn it in a way that makes sense for you.

About the understanding versus memorization debate: I know that generally speaking, it is best for students to understand the concepts, rather than simply memorize them, especially at the higher levels, but even at the elementary level. However, a familiarity with basic facts can go a long way.

I have worked with several high school students who were rusty on their basic arithmetic skills, and it significantly slowed them down. As I have said before, math is cumulative, in that the concepts build upon each other. If a student is missing earlier skills, it can negatively affect their mindset and mood about math in general. This then affects their level of learning of the later material. In other words, gaps in the material can cause gaps in the mindset and mood, which can lead to further gaps in the material.

If a student has gaps in the earlier material, it is vital not to say things that might come across as shaming. In going back to the sections on I-messages versus you-messages and intent versus impact, the ways in which parents and teachers communicate with students about gaps in earlier knowledge can be vital. It is especially important to take care not to shame a student for being "behind," or for "missing" something. The word "should" is also shaming. As an example, "you should have learned this in third grade" can crush a sixth grader's self-esteem faster than she can remedy her gaps in knowledge of times tables.

Insight #25

Access to a calculator does not replace skills in estimation, problem solving, and critical thinking.

ONE OF MY BIGGEST pet peeves is students who think to themselves, "I don't need to study math. I have this expensive gadget here. I'll just punch in the numbers, and it'll spit back the answer."

I hate to break it to you, but that's not how it works. Calculators are a tool, not a crutch. Calculators are not all bad, but they need to be used wisely. In some cases, it's best not to use a calculator at all.

Back in 2006, I coined the term "calculator abuse." In other words, students who think that punching the numbers into a calculator will yield the correct answer with little to no effort. As we discussed in the last section, math is about so much more than just numbers, or even getting the "correct" answer. It is about critical thinking, problem-solving, and finding and discovering patterns. In other words, the use of the human brain as an intellectual activity.

In 2016, the SAT test was re-vamped to emphasize critical thinking more. With that makeover, there was the addition of a math section that prohibited the use of a calculator (except in certain cases with accommodations for learning disabilities). As I predicted, a lot of students and parents were not happy about this change.

The way I see it, this change is a great opportunity to develop confidence in one's calculation skills, as well as estimation skills. After all, in a lot of real-world math, we do not need an exact answer, and sometimes an estimate is more appropriate. In these cases, estimation is a vital skill to have. On multiple-choice tests such as the SAT and ACT, it is a good skill to have to rapidly see if an answer makes sense. Because a lot of what is missing from current math education is number sense.

Besides, over-dependence on calculators can hinder self-confidence in math skills. I once had a pre-calculus student who insisted on using her calculator for things as simple as 180 divided by 30. One day, inspiration hit, so I hid the calculator on purpose and showed her that it becomes a lot simpler when you cross out the zeros to make 18 divided by 3. Then she was able to deduce the answer through a few tries without using a calculator.

Calculators can often become a source of frustration. For example, when typing equations into a calculator, you need to be mindful of the order of operations, and sometimes the parentheses can get a little tricky. Doing these simple manipulations by hand can save lots of time and energy, as well as mental and emotional energy.

This is not to say that calculators are not useful. Graphing calculators can be wonderful for producing graphs, or for doing calculations that would be tedious by hand. In addition, calculators are a useful tool for checking that your final answer is correct, or that it makes sense.

Insight #26

Word problems are especially challenging because they incorporate both reading skills and math skills.

MANY STUDENTS AND PARENTS come to me with concerns about word problems. Virtually everyone is familiar with word problems in math, such as "Mr. Smith has X apples, and he gives 5 to Mrs. Smith. If Mr. Smith has 7 apples left, how many did he originally have?" This is just a hypothetical example (and you don't have to solve it!), but you get the gist of it. Word problems require you to use math in the context of a story or a real-life situation.

Unfortunately, word problems often get a bad reputation in math, especially with Common Core. In recent years, standards have begun to emphasize critical thinking and math in context, so word problems are becoming more common, especially on standardized tests such as the SAT. So why do so many students and parents hate these word problems?

Part of it is because word problems require not only math skills, but also reading skills. Although there is usually a reading comprehension section on most standardized tests, reading skills are critical to solving word problems in math. That is because word problems require the student to understand what he or she is given, and what he or she is being asked to solve for. In other words, instead of being straightforward, most word problems require students to "read between the lines" and make inferences.

A common reason why students miss word problems is that they fail to read the question carefully. In other words, they sometimes dive into the calculations before they finish reading the question. Often, students don't check to make sure that their answers make sense. For example, would a car cost $10? Would a puppy weigh 1000 pounds? Would you have a negative number of apples? Not likely, for all of the above. That's another thing that word problems test: estimation and number sense.

For all of the above reasons and more, word problems are a common source of math anxiety. Then when they are marked wrong, they become a source of math shame, and of telling the story, "I'm bad at word problems." In that case, do what I already recommended for math anxiety and math shame: take a deep breath, and challenge the old story. Then, work on writing a new story, and start by reading the word problems carefully. Once again, word problems, like any type of math problems, take practice. But before you can start practicing, you need to develop a healthy mindset.

This will be addressed more thoroughly in *99 More Insights (and a Calculation's Still Not One!)*, releasing in 2019, but it is worth mentioning now that students with learning and reading disabilities (for example, dyslexia or dyscalculia) might find word problems to be especially challenging. In this case, all of the above applies, but also apply for special needs accommodations (if any) early. See your teacher or guidance counselor for more info.

Insight #27

Anxiety about chemistry and physics is almost always rooted in anxiety about math.

SINCE THIS IS A BOOK ABOUT STEM, it would only be right to cover the engineering and science fields, in addition to math. It is common knowledge that the majority of careers in science, engineering, and medicine require some level of math. In fact, virtually all pre-medical, dental, and veterinary programs require chemistry (both general and organic), and more often than not, require physics and calculus as well. In addition, most programs in engineering and architecture require physics, as well as calculus. Even at many high schools, chemistry and/or physics are required to graduate, in addition to at least second-year algebra in math.

Anxiety about science classes is just as rampant as anxiety about math classes. It goes without saying that math skills are an important part of success in chemistry and physics. This often means that pre-existing negative emotions about math can become negative emotions about science by association. For example, if you have had negative feelings about math since childhood, you might dread a chemistry or physics class in high school or college. Anxiety about math can frequently become anxiety about chemistry and physics and can affect success in chemistry and physics not only because of weak skills in the math material but also because of the negative thoughts and feelings at the lower levels of the pyramid. In addition, especially at the college level, these advanced science classes often have math prerequisites.

Science concepts can often be taught independently of the math skills, especially at the elementary school and middle school levels. This introduces children to the concepts of these fields without being intimidating. In addition, many high schools are breaking from the traditional mold of teaching biology, chemistry, and physics in that order, and are starting with conceptual physics before biology and chemistry. This makes sense not only in light of removing physics from math anxiety but also in terms of introducing physics concepts before chemistry. While there is a great deal of math in chemistry, most of the basic concepts of chemistry can be boiled down to physics.

My physics professor in college once said that physics is about 90 percent concepts, and about 10 percent calculations. To support this principle, he did many demonstrations for our class, including a ramp and a ball, a pendulum, and a mass on a spring. In addition, my chemistry professor in college was very sensitive to the fact that many students, especially young women, entered her introductory chemistry class with serious anxiety about math. With that in mind, she took extra time to focus on problem-solving and really laid it out step-by-step.

All in all, don't let math be an obstacle to your goals in chemistry and physics. Unfortunately, with the way it is taught and tested, it often is, but separating your feelings about math from your feelings about chemistry or physics is a good first step.

Insight #28

Calculus and trigonometry are usually not necessary for non-STEM careers. These classes keep the option of STEM careers open.

FOR MANY PARENTS who solicit my services, the ultimate goal is calculus. That is widely seen as the "holy grail" of math education, and the goal for every student to reach before graduating high school. Many parents are anxious about their teens preparing for calculus, even as early as middle school. By way of traditional reasoning, if a student is not enrolled in algebra by eighth grade, then they won't be ready for calculus by their senior year of high school. That's a lot of pressure not only for a middle school student but also for their parents.

Once again, this might sound a bit irreverent and unorthodox coming from a math professional, but for most people, trigonometry and calculus are not necessary. After all, people don't usually use calculus or trigonometry in their everyday lives. Conversely, if a person wants to pursue a STEM career, then calculus is almost always a prerequisite. According to the required courses at most colleges, virtually all STEM careers, including medicine and engineering, require calculus as a prerequisite. In addition, many advanced physics and chemistry classes require advanced math.

Opting not to take calculus will limit your chances of pursuing a STEM career. On the other hand, just because you don't take calculus in high school, or even in college, does not mean that you can't pick it up later. Although as a word of caution, it might be more difficult to pick up later, especially if you abandon math for years at a time. This illustrates the "use it or lose it" principle, which does not just apply to math. For example, I learned the basics of the piano as a child and then didn't play at all for years. I recently picked it up again as an adult and had to start at the beginning of the piano curriculum, re-playing some of the simple songs that I learned in second grade. Thus, this illustrates the "use it or lose it" principle.

Even I'm not immune to the "use it or lose it" principle in math. A few years ago, I was asked to tutor a college student in differential equations, and on my lunch break, I frantically looked at my old books from college, because I hadn't taken the course since 2002! Although the material came back quickly, it took some effort, because I hadn't seen it in fourteen years. If you or your student is considering taking a break from studying math, it would be wise to consider it carefully, because it might be more challenging to pick back up again years later.

If you are positive that your passion is outside of STEM, then it is generally okay to skip calculus. Then again, think about your reasons for eschewing calculus. If it is because of anxiety, it could easily become avoidance, and then spiral into the AAS Triangle (remember that?). On the other hand, if it is because you are genuinely not interested, then you can feel free to skip calculus.

Insight #29

Statistics, personal finances, and applied math can make math seem less intimidating, and more relevant.

MOST HIGH SCHOOLS require at least second year algebra to graduate, and most colleges require at least second year algebra for admission. Many schools require (or at least strongly recommend) additional math courses. As we discussed in the last section, while trigonometry and calculus are almost always required for STEM careers, they are often optional for those who wish to pursue fields other than STEM. This does not mean that these people should not take additional math classes. In particular, alternatives such as statistics, personal finance, and applied math can make math seem relevant to their everyday lives and can be a good alternative (or supplement) to trigonometry and calculus.

The fact of the matter is that most high schools are sorely lacking in the education of personal finance. In other words, most high schools do not offer courses on topics such as taxes, credit cards, mortgages, investments, and other related topics. This is unfortunate because these topics are much more relevant to our everyday lives than trigonometry or calculus. Many people avoid these classes because they involve math, and because they were already traumatized by math shame in the past. That is also unfortunate not only because these topics are relevant to their daily lives, but also because these classes can be a good way to make math less intimidating.

For example, in personal finance, everyone has to deal with credit cards and taxes. If you don't understand where your money is going or where it's coming from, then it's going to be in chaos, and you might not make wise financial decisions. Speaking of unwise financial decisions, a lot of young people in their late teens and early twenties get into trouble with credit cards because they don't understand how credit works. Lastly, when making a major purchase such as a home or a car, it pays to comparison shop, and to get the best deal. Unfortunately, a lot of people who abused credit cards in their 18-24 years could be paying the price for many years to come, because of a compromised credit score.

Similarly, many adults today do not understand the basics of statistical analysis, or even of reading a graph or a bar chart in an article in the newspaper or online. This especially came into play during the 2016 presidential election, when many statistics were skewed and manipulated and were not presented in a statistically sound manner. Ditto for global warming and climate change. With people uneducated, not only in the scientific principles behind these phenomena but also the mathematical phenomena used to describe them, it is no wonder there is so much denial when it comes to climate change. Therefore, statistical literacy can literally save our planet.

It should be cautioned that courses such as statistics should not be viewed as "watered-down" alternatives to calculus. Statistics and calculus both have their place in the world, and part of the beauty of mathematics is understanding the role that each one plays.

Insight # 30

Organization and executive functioning are critical skills for success in math.

WHEN IT COMES TO organization, the most important thing is to find a way that works for you. Whereas some people might work best at a tidy desk, others might work best in a less formal setting, perhaps sprawled on their bed with their laptop and with music playing. I always say that if you find a method of organization that works for you, then stick with it. No need to fix it if it isn't broke. I know this from personal experience, because many times, people have tried to force their methods of organization on me. For example, in high school, several teachers enforced what they called "double column" notes, in which you would take notes on one side of the paper, and then fold it over and write a question about the notes on the other side. While it worked for some students, it didn't work for me. I liked having room to spread out over the entire paper, and with me being left-handed, I often swapped the left and the right side. I also liked to use bright colors to color-code things (by now, you probably know that I'm obsessed with anything purple).

This is not to say that organization is not important, and this doesn't just apply to math. In cooking, there is a French term called *mise en place*, which literally means "everything in place." Before you begin cooking, read through the recipe, and have all your ingredients and equipment handy. The same thing can apply to math, or studying any other subject. Before you start studying math, make sure you have everything you need. In my years of teaching, I cannot tell you how many students forget things as basic as a pencil. That forgetfulness adds up, and if it becomes a pattern, it can really slow you down.

In addition, while I will discuss ADHD and other learning challenges at length in my next book, releasing in 2019, it is worth mentioning now that a common theme is a lack of executive functioning, which can also affect performance in math. What is executive functioning? Like its name implies, it is like the CEO or the boss of the brain. In other words, organization, time management, and keeping track of responsibilities. Executive functioning is often a challenge for kids and teens (and adults!) with ADHD and related challenges.

Another word about organization: a lot of times, what parents and teachers might do by hand, is done electronically for today's teenagers and young adults. For example, writing reminders to oneself, or keeping track of assignments or test dates. Once again, if you've found a system that works for you, then by all mean, stick to it. For example, my friend, who's only 10 years older than me, swears by her paper and pencil system and has cabinets full of hand-written notes in her at-home office. I, on the other hand, do virtually everything on my iPhone and laptop. As I always say, if you find a system of organization that works for you, then all the more power to you!

Insight #31

Math is as much a part of a well-rounded education as history, literature, and the arts.

MOST PEOPLE, AND EVEN most educators, tend to separate math and science from the humanities. In other words, while they readily see the overlaps between history, literature, the arts, and even the social sciences (i.e., psychology, philosophy), they rarely see how these apply to math and science, or vice versa. Math and science are as much a part of human creation as are literature and the arts.

When I went to Scripps College in Claremont, they had a signature "core" curriculum in the interdisciplinary humanities that all first-year students were required to take. It was a semester-long course taught by approximately 15 professors of varying disciplines, in which they would take turns giving lectures, and assigning readings, films, and other supplemental materials. We would then break into groups for our weekly discussions. My favorite calculus professor had a signature talk that he gave on the history of mathematics, and he was an instant hit, even among the students who described themselves as "not math people." (Remember, there's no such thing as a "math person.")

Part of his talent was his ability to connect math to the history of humanity, and to show how math has impacted people across the world, and different cultures. Remember, math is a human endeavor, and Pythagoras contributed as much to society as did Shakespeare or Van Gogh. Math and science are influenced by the society and culture in which they take place, and vice versa.

Just as well, math promotes critical thinking skills, reasoning skills, and problem-solving skills, all of which this professor emphasized in his calculus classes. If anything, critical thinking is what needs to be taught the most in all subjects, and my college's core curriculum definitely emphasized critical thinking. In other words, how did you come up with that answer? Are there any other alternate answers? Why is your answer the best one? And most importantly, what does your answer mean?

All too often, we think of math and science as being in a vacuum, and being an esoteric subject for a few lucky people to understand. Anyone can understand the beauty of mathematics, just like anyone can appreciate the beauty of art or literature. I'm glad that the culture is slightly changing in favor of this viewpoint. I recently visited the Museum of Mathematics in New York, and they had several exhibits of the history of mathematics, as well as integrating mathematics into the arts. For example, they had a probability simulator, and they also had a large version of the Spirograph.

Remember the Spirograph? You might have played with it when you were a kid, to make colorful spiral designs. Turns out, there's a lot of math behind the intricacy of those designs. Ditto for their exhibit with the square-wheeled bicycle, and the calculations used to design the precise path that would fit this bicycle. As we can see, math is a vital part of human history and culture.

ROADBLOCKS TO SUCCESS

(Insights 32-44)

Insight #32

Until you uncover your negative thoughts and feelings about math, your attempts to study the material will not be as effective.

WHEN THE PROBLEM is poor grades and/or test scores in math, most parents and teachers (and even students) assume that the solution is to work more on studying the material. In other words, "buckle down," and "study harder." Phrases such as those are rarely effective because there's a whole other layer that needs to be dealt with before we can most effectively work with the material.

Remember the 7M Pyramid? To refresh your memory, the base of the pyramid includes the mindset, mood, and the two spiritual levels (meditation and manifestation). And the spiritual levels are optional, but are highly recommended.

The mindset and the mood are crucial and are unfortunately a missing ingredient in much of math education today. Even in one-on-one education (aka tutoring), most of the focus is on the material, and possibly the method as well. This is rarely effective long-term unless a negative mindset and mood is addressed.

This is how I'm unique, and different from most other math tutors. A typical math tutoring session begins with the student walking in with his or her textbooks and calculators, and announcing "we have these problems for homework," or "I missed these problems on the test." Then we get to work correcting the ones they missed and focusing on their mistakes. After all, that's the way I operated for over a decade. But during that time, I realized that something was missing. By focusing on mistakes, we focus on what's lacking. I now take a different approach, in that I focus on what's already there. In other words, you are already whole, but we are just going to up-level. That's where the mindset and mood come in.

Before we can cover the material, we need to address the root causes of the negative mindsets and moods. For example, fears and limiting beliefs. If you think about it, today's education system is very fear-based, both for students and parents. This is unfortunate because these fears keep people stuck. When people stay stuck, it is unlikely that our society will make progress. We need to challenge these fears and limiting beliefs, and go forward. In order to go forward and understand the material, we need to go through the old negative stuff, and reframe it and release it. That is why it is vitally important to do the mindset work before starting on the material.

Even parents and teachers can have negative thoughts and feelings about math. These negative thoughts and feelings can affect students, albeit unconsciously. Parents' negative thoughts and feelings about math often form the basis of students' negative thoughts and feelings about math. That is why students and parents (and teachers, if possible) need to work together to transform these negative thoughts and beliefs.

Insight #33

The process of learning math goes beyond the material (times tables, fractions, algebra, calculus, etc.)

A FREQUENT QUESTION that I get is, "what levels of math do you tutor?"

Of course, they're expecting answers like elementary math, middle school math, algebra, geometry, trigonometry, calculus, etc. In other words, they're expecting an answer that relates to the math material. While this is a legitimate question, and a legitimate answer, it doesn't tell the whole story. My technique goes beyond the material. The material is just on the surface. In reality, we're dealing with a much deeper issue.

No matter what material we're dealing with, we're always dealing with the method (study skills), mindset (thoughts), and mood (feelings) around math. While some material might be more challenging than other material and there is no question that the material on math builds on itself (i.e., it is cumulative), we can't ignore the thoughts and feelings that come up as we're doing the material.

When a parent or teacher asks me the above question, I answer that I tutor all levels, from elementary school through advanced calculus and differential equations (usually taught in college). I add the caveat that learning math goes well beyond the material. Most parents, and even most teachers, are not aware of this. They think that learning math is just about the material. I invite you to take a deeper look at this. There are so many layers to learning math that require a non-academic, spiritual approach. That is where the current academic system is not balanced, and what leaves the students and parents to suffer.

In going back to the house analogy, remember how we couldn't decorate, or even furnish, the house without a solid foundation? Same principle here. It doesn't matter what kind of decorations or furniture we want for the finished house. The foundation is always going to look basically the same but is going to be unique for each house. Nevertheless, the foundation is always going to be the most fundamental piece.

Your foundation is your thoughts and feelings about math. Once we have addressed your thoughts and feelings about math, we can go back and add the furniture and decorations (aka the material). What good would furniture and decorations do without a solid foundation? Think of that the next time you want to fix the material before looking at your thoughts and feelings.

When I say the term "math foundation," many people tend to think of elementary-level math (i.e., times tables, long division, fractions, decimals, etc.). Once again, while these concepts are vitally important, and are critical skills for algebra and calculus, they are not nearly as foundational as the thoughts and feelings about math. After all, if you have limiting beliefs about your math ability, how can you master any of the material? This even applies to the elementary-level material, and even if you are an adult. Playing "catch-up" can cause a not-so-healthy mindset and mood. Because of this, it is vitally important to be aware of these negative thoughts and feelings and to nip them in the bud as they come up.

Insight #34

Math is not just numbers. It exists in a social and emotional context.

WHEN MOST PEOPLE think of math, they think that it is completely linear, logical, and objective, and is devoid of any emotions. After all, in math, there is usually a right or wrong answer and does not have as many subtleties and avenues of interpretation as does art, literature, or history.

We do need to integrate social and emotional development into the learning of math. Although math is commonly seen as an objective subject, it brings up plenty of emotions. These emotions are not always rational. Like we said in the sections about the biological basis of behavior, if you try to rationalize your way out of these irrational emotions, you only make it worse. Think of the main characters in *The Wizard of Oz*. The Scarecrow was all about the brains, the Tinman was all about the heart, and the Cowardly Lion was all about courage. Most people think of math as a "Scarecrow" subject, in that they think it involves brains only. We need to integrate just as much heart and courage into learning math.

People often say to me, "I was never good with numbers." Well, guess what? Math is about more than just numbers. It is about the context of these numbers and fitting them into patterns. Really, math is the study of patterns. Even young children make patterns in their lives. Sadly, once people see math as this esoteric subject, it intimidates them, and they shy away. This is where the emotions get in the way. Ditto for the social context.

Remember our findings about middle school? Many girls tend to lose interest in STEM during middle school due to social pressures and intensifying gender roles. Many times, girls do not support each other when they are interested in STEM fields, or their mothers don't support them.

Especially at the middle school level, social support is critical, both from peers and adults.

People often see being emotional as being a weakness. Rather, being emotional can be a great strength. Emotional awareness is a gift, and we need more emotionally aware people in the math and science fields. Because after all, STEM is a human endeavor and has widespread social effects. This is why we need to bring this social and emotional awareness to the classroom, and to math learning outside of the classroom.

Just as well, I am going to challenge the common belief that it is a sign of emotional immaturity to show any sign of intense emotions. Most of the time, math teachers brush emotions aside, and often, parents do as well. But sweeping them under the rug can only make them more intense. We need to deal with them as they come. That is why my approach is unique. I integrate the "head" aspects of math with the "heart" aspects, as well as the "courage" aspects. Emotional responses to math are welcomed, and we explore and heal them. Because remember, being emotional is not a weakness. Rather, it is a strength.

Insights #35-37

#35 - *There is a significant difference between a challenge and a struggle.*

#36 - *If you assume that your children are going to struggle, then their confidence will decrease.*

#37 - *Avoiding challenges can backfire.*

THERE IS A MAJOR difference between a challenge and a struggle. While the words are often used interchangeably, the energy behind each one is very different. Struggling implies that you're stuck and that there's no way out. Imagine the visual of a person drowning, frantically flapping their arms, and the waves crash over them faster than they can flap or scream. In other words, feeling overwhelmed, or engulfed. On the other hand, a challenge is something that you have the strength to overcome and will make you stronger and grow your character.

Struggle focuses on what's going wrong, challenge focuses on your inner strength. In other words, the difference lies in your power, and your personal agency. A struggle implies that the obstacles are too big for you to overcome, but a challenge implies that it is not only doable but also character-building.

On the other hand, "easy" or "with ease" implies that there are no challenges and that you don't have to put in the effort. We want to discourage that for several reasons. For example, many people have suggested that I advertise my services as "math with ease." I believe that does a disservice, both to parents and to students. That is because it undervalues the process of effort. After all, nothing worthwhile comes without effort. Effort is not synonymous with suffering, and that's where we must draw the line between a struggle and a challenge. It's kind of like the Goldilocks principle. Remember Goldilocks, who complained that one porridge was too hot and one was too cold, and the other was just right? Same thing here. Whereas ease is too little effort, and a struggle is so much effort that it makes you suffer, a happy medium is a challenge. It's kind of like the model of the comfort zone, the stretch zone, and the panic zone.

In this analogy, the comfort zone represents ease. While our comfort zone is familiar, we do not want to stay in our comfort zone forever, because we do not grow or change in comfort zones. Many people like to stay in their comfort zone, and many parents and teachers hope that my services will put math in one's "comfort zone." Once again, I feel that this would be a disservice. That is because many people who avoid math do so because they unconsciously want to stay in their comfort zones.

Math is in the "panic zone" for many people. Not just students, but also parents and teachers as well. You know the feeling. Math causes anxiety, tension, and self-doubt and limiting beliefs for many people, especially women. That is the panic zone, which we want to avoid as well because it feeds into the AAS Triangle of anxiety, avoidance, and shame.

The happy medium is the stretch zone, which is slight discomfort, but major rewards for this slight discomfort. As uncomfortable as it may be, let's talk about discomfort. Paradoxically speaking, discomfort helps us to grow, and avoiding discomfort will help keep us stuck. Naturally, we want to avoid discomfort, but stepping out of our comfort zone is a major step. This can also apply to adults who have stayed in their comfort zone by avoiding math for years (or decades!).

It all boils down to your thoughts about math. When you have negative thoughts, it is a struggle. When you have positive thoughts, it is a challenge. This speaks to the importance of mindset. Remember the 7M Pyramid?

This has to do with shifting our thoughts, and in turn, shifting our feeling. This is the main difference between a struggle and a challenge.

Many parents are scarred by their own past experiences with struggles and want to protect their children from struggling. Parents might discourage their daughters from pursuing STEM, in the interest of protecting them from the negative thoughts and feelings that they went through. If a parent assumes that a child will struggle, then the child's self-esteem will suffer. In other words, when a parent or teacher assumes that a child will struggle, it reflects their lack of confidence in the child.

I am all too familiar with this phenomenon. Not in math, because math almost always came easily to me. In terms of my challenges with social situations and my weight and body image, many well-meaning friends and family members have assumed that I would struggle in these areas. That affected my confidence in these areas, because I assumed that attempting to change them would cause me to suffer. Many of the parents that I have worked with in the past, albeit well-meaning, have reflected this sentiment on to their daughters (and sons!) when it comes to math.

On the other hand, avoiding challenges can backfire. Many people tend to confuse struggles with challenges, and they associate struggles with suffering. They avoid challenges as an unconscious way to protect themselves from suffering. There will be some discomfort and pain, but it will be a great opportunity for a growing experience. Many students avoid challenges in math, and many parents subtly encourage their children to avoid challenges in math. This is especially true for girls. So how do we remedy this? Encourage girls to take on the challenge, and when they stumble upon roadblocks, help them to see it as a temporary detour, and not as a permanent roadblock.

Part of the paradox is becoming comfortable with discomfort and resisting the urge to immediately "fix" any sign of discomfort. This once again goes back to the limiting beliefs of the "tutor mentality." All too often, we feel the need to have someone come and wave and magic wand and "fix" grades that are not up to our standards. It rarely works that simply, and we need to be okay with the complexity. And not only okay with it, but also to embrace it. Because it's not just math that's like this. Life is like this.

It brings me to the Martin Luther King Jr. quote, "we must accept finite disappointment, but never lose infinite hope." In other words, one disappointing grade or other experience is not the end of the world. On a similar note, Gabby Bernstein's quote, "infinite patience leads to immediate results." In other words, be kind and gentle with yourself. This goes for both students and parents.

Our current educational system takes a very linear approach. In going back to the theme of challenges versus struggles, remember that the educational system is not the end of the world. In other words, one grade is not going to dictate your entire future. Sure, it might have consequences, but you can live with those consequences. Nothing rewarding ever came from staying in one's comfort zone. Just as well, the mindset of struggling is going to keep you from moving forward and is going to keep you locked in a battle. All too often, when parents hire me to tutor their students, the students and parents are engaged in a war. In other words, the struggle looks like a war, with one battle after another. Over time, these battles can add up.

Think of challenges as character-building exercises. In a popular children's book from Mexican culture, *Arrow to the Sun*, the hero has to pass through several tests, or "kivas" before he can receive his treasure. When I first read this book when I was in kindergarten, I got shivers from the "kiva of lightning." In spite of the scary scenes, what I remember most about the book was how the main character changed for the better as a result of the kiva of lightning. So maybe you can see this as the kiva of algebra, or the kiva of geometry. Or more importantly, the kiva of math mindset, regardless of what level material you are studying (or even if you haven't studied the math material in years).

If you're in doubt about whether or not you should pursue a challenging experience, then go ahead and do it! Be aware of your resistance as it comes up, and reframe it. And remember the difference between a challenge and a struggle, and if things get challenging, remember to breathe. You've got this!

Insight #38

Most math tutors prioritize grades first, feelings second. The Math Lady reverses these priorities.

ONE OF MY FAVORITE memes on social media states, "your child's mental health is more important than their grades." Too true. Many parents and teachers seem to forget this, especially in our test-driven culture.

While the reality of the situation is that grades are important, at least in the short run, there are many more important things. In other words, we do not want our students to get good grades at the expense of their mental health. So how can we find a balance between achievement and mental health? As paradoxical as it may sound, greater achievement in math (or any other subject) occurs when emotional well-being is improved. In other words, emotional well-being comes first, and then the achievement follows, and not the other way around. Unfortunately, our culture's emphasis on grades and test scores can be detrimental to mental health, both for students and parents.

Emotional well-being is connected to physical well-being and spiritual well-being. That is why I include those components in my coaching, along with the academic components. Although many people might not think those components necessary, and might even argue that they take time away from the academic work, they are vitally important to increasing achievement and confidence in math.

Today's approach to math is quite unbalanced. In my approach, we have four components to successfully developing skills and confidence in math (or any other subject): academic, emotional/spiritual, creative, and social/

political. Most math tutors are strictly focused on the academic, and ignore the other three aspects. They brush aside the feelings that come up with math. All too often, when a student is working on math, some uncomfortable emotions can come up. Sometimes, they might even cry.

Many parents and teachers and tutors are uncomfortable with tears. They might say, "Are you okay? Don't cry!" and then jump into problem-solving. I firmly believe that you need to feel it to heal it. With that in mind, I prioritize healing before problem-solving. But sometimes, they can take place simultaneously. More importantly, it is vital to find a balance between healing and problem-solving.

In going back to the 7M Pyramid, the levels of mindset and mood are vitally important. You can't have thoughts without feelings, and vice versa. In boiling it down, feelings have more impact than thoughts. In other words, you can know something rationally, but still not feel it emotionally. For example, you might know in your head that competition is unhealthy, but you still can't help comparing yourself to others. In this case, it is best to acknowledge the feelings, but then choose to feel a different feeling.

Don't skip this step, because if you don't acknowledge the feelings, then they will fester and snowball. The same is true for parents. You need to be aware of your feelings and shift them in order to best support your children.

Insight #39

Holistic health looks at the whole person, not just the symptoms. Same principle with holistic math.

I FIRST GOT THE IDEA for the term "holistic math" as I was driving along the highway in Edmonds, WA, and saw a sign for "holistic dentistry." I was intrigued, so I looked this dentist up online. He has a unique approach that integrates all aspects of not only physical health but also mental health, into dentistry. In other words, with him, a cavity is not just a cavity, and a root canal is not just a root canal. There is usually an underlying cause, and the health of one's mouth reflects one's overall health.

You've probably heard of holistic health, and might swear by it, or might be skeptical about it. Nevertheless, as its name implies, holistic health looks at the whole body and mind, and not just the symptoms. In contrast to conventional medicine, holistic practitioners evaluate the whole person and take into account relationships, nutrition, stress, and environment, among other things. In other words, they don't just look at symptoms, because symptoms rarely occur in isolation.

In contrast, most conventional doctors (and dentists) focus exclusively on the symptoms. For example, if somebody comes in with a cavity, you fill it. If someone comes in with a sore throat, you give them throat lozenges, or antibiotics if appropriate. A lot of times, these "fixes" just mask the symptoms and don't address the root cause. For example, taking a pill for a headache does not address the root cause of the headaches, which could be anything from dehydration to a hormonal imbalance.

Similarly, when it comes to math (or any other academic subject), most tutors just look at the "symptom" of low grades and test scores and attempt to remedy it by repeatedly going over the material. This rarely addresses the root cause of math anxiety and math shame, which are rooted in mindset and mood, and even in the spiritual levels. Thoughts and feelings are vitally important to success in math, just like they are to health.

Here, we look at the whole picture. This includes physical, mental, emotional, and spiritual, not just academic. Instead of going straight to the academic "Band-Aid" fix, we take a look at your overall goals, your past experiences, your mindset, and your fears, among other things. In other words, we address the root cause of the symptom of low grades and test scores.

Just like with holistic health, we take an individualized approach. In other words, no two cases are alike, and each one must be evaluated individually. For example, the root cause of one student's anxiety might be completely different from the root cause of another person's anxiety. Just like with two patients presenting with the same symptoms in a health setting, we need to get to the root of each one individually, and not prescribe a one-size-fits-all approach. In going deeper, we ensure a longer-term approach and not just a quick fix.

Insight #40

Dieting is often a "Band-Aid" quick fix for body shame. Same principle with extra worksheets and math shame.

I AM A VETERAN of many years of dieting. Weight Watchers, Atkins, the Zone, calorie counting, carb counting, clean eating. You name it, I've done it. Similarly, many of you are probably veterans of many years of tutoring. You might have tried several standardized test prep centers, online programs, private tutors, and after-school programs, and nothing seems to stick. Believe me, I empathize. There are many striking parallels between my struggles with weight and my ideal reader's struggles with math. I know what it's like to feel ashamed, defective, pressuring yourself to try harder, comparing yourself to others, and getting discouraged and giving up. In other words, feeling like you can't do anything right.

What I realized was that as paradoxical as it may sound, in order for me to make peace with food and my body, I needed to let go of my attachment to the number on the scale. You might be mad at me for saying this, but I have found that the best results come when students and parents let go of their attachment to the grade or the test score. In other words, focusing on the process more than the results. Not only that, but also redefining the results.

I know that in both cases, this sounds like the complete opposite of what society says, but please hear me out. In both cases, an emotional meaning is attached to an arbitrary number. In the case of dieting, the number is the number on the scale, and in the case of tutoring, the number is the grade or test score. In both cases, the limiting belief is that this number defines you as a person, and is a reliable indicator of one's progress. This number carries a lot of emotional weight (no pun intended) and can be a trigger for deep shame.

To break this cycle, we need to avoid our quick "Band-Aid" fix. In other words, the solution to math shame is rarely to study the material harder. In fact, that can exacerbate math shame. Ditto for dieting and body shame. I found that the more I dieted, the more shame I felt about my body. Then when I felt shame about my body, I would start a new diet. It's a vicious cycle, and it doesn't get you anywhere.

In both cases, we need to go deeper. Explore your shame from the past, and let it go. Then, think about what results you want. There's nothing wrong with wanting to improve your grades or test scores, just like there's nothing wrong with wanting to lose weight. But let's take a deeper look at these goals: what is the feeling behind these goals? Would you be OK if you got that feeling, but didn't get the grade or test score (or the number on the scale) that you wanted?

My friend once had a goal to lose 50 pounds by her 40th birthday. She swapped fast food for a meal kit delivery service and downloaded the Couch to 5K running program on her iPhone. As she got into her running, her focus shifted. Finally, shortly after her 40th birthday, she ran a half-marathon, even though she was still technically "obese" by BMI standards. This story goes to show that the number on the scale does not tell the whole story and that there are other ways to measure results. We can then apply this to math as well, once we shift our emotional attachment to grades and test scores.

Insight #41

It is okay to progress through math at your own pace.

MANY PARENTS AND teachers reach out to me because they think that their children and students are "behind" when it comes to math. Once again, they are adhering to the age-based, lock-step model that assumes that if a teenager does not graduate high school at 18 and graduate college at 22, then they will be a failure for the rest of their lives.

However, it does not have to be that way. Opportunities are circular and can occur at any age. The feeling of being "behind" promotes a sense of pressure, which can trigger anxiety, and can also trigger shame. The idea of being "behind" compares students to not only an age-based standard, but also compares them to others, which might trigger insecurity, not only in the students but also in the parents. Parents might see students being "behind" as a poor reflection of their parenting, which might trigger feelings in the parents.

When it comes to life, as well as math, there is no such thing as being "behind." Everyone has their own unique timetable and their own special rhythm. If you try to force children to conform to a certain timetable, then they will likely shut down, and go into the AAS Cycle. In most cases, it is best to honor one's own special rhythms.

The idea of grouping children by age has come into question in recent years. Although intellectual (and social and emotional) milestones might be correlated with age, correlation does not imply causation. In other words, in a group of children or teenagers of any given age, it is to be expected that there will be a wide variety of abilities and skills. The lock-step age-based paradigm hurts not only those who are "behind," but also those who are gifted and might require more challenging work.

While math is generally cumulative, some concepts can be introduced out of order. For example, there was once an article circulating on the internet about how even some kindergarteners could understand the basic principles of calculus. In case you're curious, at its most basic, calculus is the study of rates of change. In other words, how fast does something change? Taking that concept, they found that even some 5- and 6-year olds were able to understand rates of change at a fundamental level. Children often know more than we give them credit for, especially when those concepts are not formally tested.

As paradoxical as it may seem, sometimes, slowing down the pace of the material can increase the rate of understanding. Sometimes, all it takes is one "a-ha!" moment (to quote Oprah) for things to make sense. Going at a pace that's faster than a child is comfortable with can sometimes make the mindset and mood worse, not better. In other words, while the material is being presented, we sometimes need time for the mindset and mood to catch up. In other words, meeting a child or a teen where they are (instead of where they "should" be) can work wonders for their confidence and understanding in math.

Insight #42

The red "X" on a math test or assignment can be a trigger for math shame.

IMAGINE THIS SCENE: you've studied hard for a math test, and put many hours into it. You thought you aced it, but there were a few problems you weren't sure of, so you just scribbled something down or left it blank. The day comes when the teacher hands back the test. Then, you sink in your chair. What just happened? You saw the notorious red "X" scribbled across one or more of those problems, and you felt defeated. That mark is a universal sign of shame. You then fold the test over, stuff it into your backpack, and sulk. Unfortunately, this happens all too often.

Why is that? Because of the emotional undertones beneath the red X. Many adults I've talked to have recalled vivid stories of tests that were covered with red X's, and it was all along the same theme: feeling stupid, mediocre, and believing that one wasn't working up to their potential. Needless to say, the red X is a shame trigger for many children and teenagers, and even lingers in the hearts and minds of some adults years later. As a reminder, Brene Brown defines a very basic difference between guilt and shame: guilt means you've done something wrong, and shame means you are something wrong. In other words, shame reflects what is perceived to be a defect of character, and a flaw as a human being.

More recently, in my former job, we would use an online learning program that included online tests. In these online tests, the questions had to be answered in the order in which they were presented, with no skipping and coming back to difficult questions. What's more, these tests would give immediate feedback, and although it looked slightly different than the hand-written red X's of yesteryear, they would mark their "incorrect" responses with a red X. In a sense, the same mark, and the same principle of it being a shame trigger. I would watch with pain and compassion as a student clicked on a response that was close to being correct, only to be met with the red X. This was particularly painful if it was towards the beginning of the test, because that red X would inevitably affect their mood. Because the red X was a shame trigger, that feeling would haunt them for the rest of the test. That is why I firmly believe that immediate feedback is psychologically harmful, especially if it can be a shame trigger.

This does not mean that students (or parents) should avoid shame triggers. Instead, become curious about them. And not just focusing on the material that you "missed" on the test. How does this red X make you feel? What memories does it bring up?

Insight #43

The upsides of perfectionism can include high standards and attention to detail.

PERFECTIONISM IS OFTEN, but not always, rooted in shame. I've always liked to joke that I'm a perfectionist because I'm a Virgo, but I know it's a lot deeper than that. In other words, a lot of my perfectionism is rooted in my childhood shame, not about math, but about other areas, such as body confidence and social confidence. Perfectionism can manifest itself in math and can feed into math anxiety and math shame. Perfectionism tends to discount or brush aside the things the person does well, and laser focuses on the things they think need to be "fixed."

A good example is a student or parent who gets upset about one B+ while ignoring all the other A's. Or on a similar note, if you got an 80 or a 90 on a test, beating yourself up for the questions you got wrong instead of giving yourself credit for the ones you did well. Or for an adult example, if you go in for your performance review at work, and your boss gives you overall praise, but she mentions one thing that you could improve. The key word here is the word improve. We all have areas in which we can improve, and learn and grow. The key here also is being kind and gentle to yourself during the time you are changing and growing. And some degree of perfectionism is necessary for math, and in life in general. This is especially true when you reframe perfectionism as having high standards and attention to detail, both of which are important in math.

For example, one missing negative sign can change the whole solution to a problem. Ditto for inverting digits or for "flipping" a fraction. This is why it's important to pay attention to details. Same principle with words like "not" and "except" on word problems, as well as reading comprehension questions.

Having high standards is especially important in math, especially when you're dealing with students who are women, minorities, or low-income. You want to challenge every student to do the best work they're capable of, instead of assuming that they're not capable because of their race, gender, disability, etc.

For those of you who remember the show *Desperate Housewives*, Bree Van de Kamp (briefly Hodge) was a case study in perfectionism. Although her perfectionism bordered on obsessive-compulsive and unhealthy, her character also illustrates the positive traits of high standards and attention to detail. For example, when her friend Lynette's husband Tom opened a pizzeria, Bree did a book signing for her cookbook. One of the featured recipes was for four-cheese pizza. When Bree discovered that Tom had made the pizzas with canned Parmesan, she threw out the entire batch, proclaiming that she wouldn't allow her recipes to be made with "sub-standard" ingredients, and insisted that a fresh batch be made with freshly grated Parmesan.

Where do we draw the line between striving for excellence and an unhealthy attitude? A lot of it has to do with how it makes us feel, as well as where it's coming from. For example, one of my former students came to me as a ninth grader, because after years of getting straight A's in math, she got a B+ in her first quarter of geometry. Always seeing his daughter as a star student in math, her father proclaimed, "she is not a B student in geometry," and insisted that I "fix" the problem. This made her feel labeled and put into a box, and I ended up speaking to the father and stepmother about perfectionism. It's all about finding a balance.

Insight #44

Procrastination is often a sign of anxiety and avoidance.

YOU KNOW THE DRILL. You have a test or deadline in a month. You think to yourself that you have all the time in the world, so you say, "later," because to be perfectly honest, you want to avoid this task. Then the test or deadline is in a week. You start to feel mild anxiety, but you still say "later." Then it's the night before the test or deadline, and you're in a full-blown panic. Your anxiety makes it nearly impossible for you to work or study at your best. What just happened?

Welcome to the world of procrastination. We all do it, from time to time, especially on tasks that we find unpleasant. How many of us have put off our taxes until the night of April 14th? How many of us have skipped a doctor's or dentist's appointment or an oil change? Of course, we might procrastinate on these things, but we always tell our children (or our students) not to procrastinate in their studies.

The truth is, when a child or a student procrastinates in their studies, there is usually an underlying psychological reason. This is especially true in math, but can just as easily apply to any other subject. More often than not, procrastination is a manifestation of the anxiety that leads to avoidance. This then sets up the old AAS (anxiety, avoidance, shame) Triangle that we're trying to avoid. Procrastination is often rooted in perfectionism, fear, anxiety, and avoidance. And adults can relate to this too. How many times have you delayed going to the dentist because you didn't want a lecture about flossing or delayed going to the doctor because you didn't want a lecture about weight loss? We've all been there. For some students, studying math brings up feelings of frustration, confusion, and inadequacy. Most people want to avoid those feelings, so they avoid studying.

Procrastination fits very well into the AAS Triangle. Anxiety leads to avoidance, avoidance leads to shame, and shame leads to more anxiety. Procrastination rarely results in one's best effort in terms of studying, or writing, or any other type of work or creativity. In academics, especially in a test-based setting, there are usually consequences in terms of grades. This can then become a shame trigger, whether with self-talk or with messages from parents, teachers, or other adults. The key is to break the cycle of procrastination by breaking the cycle of shame.

Action is the antidote to anxiety. Often, just getting started is all it takes. When I'm procrastinating on something, I do what I like to call the "15-minute rule." That is, just do the task I've been dreading for 15 minutes. Then, after the 15 minutes are up, I can either stop or continue. And nine times out of ten, I end up continuing.

The same can be applied to studying math. Just tell yourself you'll do one problem. Or two problems, or whatever feels most comfortable for you. If one problem seems too overwhelming, try a different one. And if the whole thing seems overwhelming, go outside and take a few deep breaths, and give yourself credit for at least getting started.

REDEFINING RESULTS, REDEFINING SUCCESS

(Insights 45-50)

Insight #45

Your goal in math is not really a grade or a test score. It is the feeling behind that grade or test score.

ABOUT 99 PERCENT of the time, when people solicit my services, the result they want is a certain grade or test score for their child or student. In going back to the 7M Pyramid, this represents the top level or mastery. In other words, what result do you want? Most of the time, that result looks like a grade or a test score. Let's look at it more closely. What does that grade or test score represent for you? What opportunities will be available as a result of this grade or test score? Upon a closer look, the grade or test score may represent feelings. I invite you to take a closer look at these feelings. How does this grade or test score relate to your bigger goals?

At the beginning of my programs, I do an exercise in goal-setting. I always do this with students, and occasionally do this with parents as well. All too often, students and parents disagree about their goals for the student, and that can cause a conflict in the parent/teen relationship. It is vitally important for parents and teens to be on the same page regarding goals. The parents and teens don't necessarily need to agree. They just need to understand where the other is coming from.

Once you've clarified your goals, you also need to clarify why those goals are important to you. Once again, this is vitally important for both students and parents. What does this goal represent? Is this goal truly yours, or is it to impress another person? Is this goal hope-based or fear-based?

According to Danielle LaPorte, soul goals are based on feelings. And yes, soul goals can include pursuing a STEM career or getting into a certain college. In other words, math can be a way to pursue your soul goals. If that is the case, then math can open the doors to a bigger opportunity. In order to succeed in math, you need to be aware of your bigger goals. Thus, you need to be aware of the feeling that the grade or test score represents.

In most cases, in adulthood, grades and test scores don't matter. At least not on the surface. But the emotional reactions to grades and test scores can linger for years. If you have painful memories of past grades or test scores, can you reframe them? This is true for parents as well. Sometimes, when a parent wants a child to get a certain grade or test score, it is a way of living vicariously through them, or of unconsciously not wanting them to suffer the way they did. Those are all noble goals, but they put the focus more on grades than on feelings. The key is to become aware of the feelings.

All in all, what does the grade or test score represent to you? If anything, grades and test scores are symbols for feelings.

Insight #46

You cannot improve your grades or test scores in math until you improve your mindset in math.

ONCE AGAIN, in going back to the 7M Pyramid, we need to look at the levels of mindset and mood. In other words, thoughts and feelings. What are your thoughts about math? What are your feelings about math? As I said before, when people start working with me, their goal is almost always a grade or a test score. And that's a reasonable goal. Like we said in the last section, the ultimate goal is not a grade or a test score, but the feeling behind that grade or test score. Thoughts and feelings are inter-connected. In other words, your thoughts affect your feelings and vice versa. Often, it's more effective to start with the thoughts. In other words, the mindset. Thus, your mindset determines your grades and not the other way around.

It is the same principle with other goals, such as money goals and health goals. In terms of money, mindset determines wealth and not the other way around. When I was discussing the concept of money mindset with a friend, and I mentioned a mutual acquaintance who had some limiting beliefs about money, my friend said, "How can she have a faulty money mindset? She's rich." This case goes to show that you can have millions sitting in the bank and still have a negative money mindset. Conversely, you can be homeless and still have a positive money mindset.

The same concept is true for math. You can be earning A's in math, and still have a faulty mindset about math. Likewise, you can be earning C's in math, and still have a healthy mindset. Once again, I know this sounds like a blasphemy coming from a math professional, but I firmly believe that it is much more beneficial in the long run to be earning C's in math with a healthy mindset than earning A's in math with an unhealthy mindset.

If your grades in math are not what you would like and you have limiting beliefs about math, there is a good chance that exploring, releasing, and reframing these limiting beliefs will help you improve your grades. The caveat is that the mindset work has to come before the studying. This is where my approach is widely criticized and misunderstood. After all, the traditional approach goes straight to studying the material and skips the mindset work. However, in order to have long-term results, you can't ignore the mindset work.

Mindset work can consist of a lot of tools, but one of the most important tools is to reframe automatic negative thoughts, or as I like to call them, "ANT's." We all have them. We all have thousands of thoughts a day, most of which we are not aware of. When you become aware of a negative thought that occurs persistently and reflexively, you develop the power to change it. And changing your thoughts is the key to changing your mindset.

Insight #47

Even if a student gets an "A" in math, the grade means nothing if the student suffered in the process.

NUMEROUS MATH TUTORING companies advertise their results as "A-plus, guaranteed!" Or something along those lines. In other words, they focus on the grades and test scores. That should not be the only focus. These places rarely, if ever, address the mindset and emotional aspects. These places often inadvertently contribute to the suffering that so many people feel in math.

When someone says to me, "I wish you would have been around when I was young!" it tells me that they suffered in math when they were younger and that they have not completely healed from that suffering. In other words, the scars from this suffering can last for years, or even decades, and can affect people for a lifetime. It is proof that math is a leading cause of suffering for people of all ages. Part of my mission is to assuage that suffering and to provide an alternative approach that avoids, or at least minimizes this suffering. Some students, parents, and teachers are so hung up on the "A" that they allow suffering to happen as a result of striving for that "A."

That is why it kills me when parents say that their kids just need to study harder. There is a big difference between studying harder and studying smarter. Part of studying smarter is to consider the whole picture, including your mental, emotional, and even spiritual well-being. Because those aspects are vital to preventing suffering. Unfortunately, the academic-based approach promotes suffering. Even if a student does get the elusive "A" (or a perfect score on a standardized test), it means nothing if the student suffered in the process. After all, it is the feeling and not the grade, that is the main motivating factor. If the feeling is not-so-great, even if the grade or test score is stellar, the negative feeling will ultimately win out, and the student will ultimately give up on STEM.

My best advice is to strive for A's in the subjects you are wildly passionate about and learn to be okay with B's for the other subjects. After all, perfectionism is a major killer of a healthy mindset and mood. In going backward, it is a healthy mindset and mood that drives achievement and mastery of the material. In other words, by taking the traditional top-down approach, you are encouraging suffering and perfectionism.

Even going into the spiritual levels, part of the work is in realizing that you are "good enough," and that grades are not a reflection of your self-worth. This often has to be done for both students and parents. In other words, parents often see their students' grades as a reflection of them as parents. Even if they don't state it directly, their students can pick up on it and can absorb that pressure. Then they internalize these beliefs, and they drive themselves too hard, and they suffer in the process.

Insight # 48

It is your student's attitude about math, not their grades in math, that will determine what STEM opportunities are ultimately available to them.

IN ADDITION TO alleviating the suffering that frequently accompanies math, another part of my mission is to increase opportunities in STEM, especially for women. Unfortunately, these opportunities are often woven into the rigid, lock-step educational system that is prevalent today. In other words, the expectation that all students will graduate high school at 18, graduate college at 22, and then go on to graduate or professional school, or enter the workforce. It is a common belief that if a student deviates from this timetable, then they are "behind," and opportunities are permanently lost.

It doesn't have to be this way. Opportunities are cyclical, and everyone has their own unique timetable. While grades are important in today's system, they do not tell the whole story. As we explored in the previous two sections, it is a student's mindset that determines their grades, and not the other way around. A big part of revolutionizing math and increasing the availability of STEM opportunities, especially for girls, lies in the mindset work.

Unfortunately, many parents and teachers are not willing to do the mindset work and many people are not aware of the importance of mindset. Then they place all the emphasis on grades. This can all contribute to a faulty mindset. It's a vicious cycle, and a "catch-22," so to speak. It is best to start at an early age on mindset work.

The grade-based system of standardized testing is a reflection of Carol Dweck's fixed mindset. In other words, the belief that a person's abilities are set in stone, and grades and test scores are a reflection of those abilities. Thus, this mindset limits a student's opportunities. The "it's too late" mindset that so many adults have is also a reflection of the fixed mindset and also limits their opportunities. The reality is, it's not too late for anyone to change their mindset about math. It is never too late to seize new opportunities in STEM. Because I'm all about increasing opportunities in STEM, regardless of age, gender, race, sexual orientation, or ability. But a faulty mindset can cost many opportunities in STEM.

A lot of those limiting mindsets are rooted in limiting beliefs about race and gender. And those limiting beliefs are often passed down through generations, from parents to their offspring. Of course, this is all done unconsciously, and parents and teachers are often well-meaning with their advice that is based on limiting beliefs. This advice can often backfire.

The younger generation is likely to internalize this negative mindset, and that limits their achievement, and therefore, their opportunities. Just as well, even if you think that an opportunity is "unrealistic," apply for it anyway. The definition of "unrealistic" is subjective, and a lot of times, people limit themselves in the name of being "realistic." In going to the spiritual levels of the pyramid, manifestation does not care whether something is realistic. It just cares whether you're aligned to something. When you're aligned to a goal, you take inspired action to make it happen.

Insight #49

Grades and test scores are not always an accurate reflection of a student's achievement or understanding in math.

WHEN PARENTS AND teachers check on the progress of their students, they're usually interested in grades and test scores. Unfortunately, grades and test scores don't tell the whole story. Many parents and teachers are not aware of that, and take the grades and test scores at face value. This can backfire in several ways.

First and foremost, if the test is not well-designed, it may not be reliable in measuring achievement in the material that it is designed to test. For example, at my former job, we used an online program that had built-in tests. The test questions not only had a lot of typos, but they also had a lot of questions that were ambiguous and could be interpreted in different ways. The computer system did not give partial credit, especially on the "choose all that apply" type of question. If you chose one correct answer, but not all, then the entire question was marked as incorrect. This program was quite discouraging, both for students and for parents.

Whether a test is done by hand or online, the test itself doesn't tell the whole story. Most parents and teachers (and students) are conditioned to believe that if a student receives a poor grade on their test, then it means that they don't understand the material. This could not be further from the truth. Not only could the test have been worded ambiguously, but there are a number of factors that go into a test score. For example, some students do not do well on timed tests. Although some places grant extended time as an accommodation for learning disabilities, this does not always work, in that no matter how much time a student has, they can have test anxiety. Remember the limbic system? No matter how well they understand the material, if the limbic system is over-active, they are likely to draw a blank on the test.

When a student does not do well on a test, it may not because he or she can't or won't understand the material, but because he or she has mental and emotional roadblocks. In going back to the pyramid, this goes beyond the material—to the method, mindset, mood, and even the spiritual levels. If a student is not in the right energy to take a test, no amount of strategy will help.

What do we do instead? Don't treat grades and test scores as the be-all, end-all. Remember, behind every grade or test score is a story. A story that deserves to be heard. Instead of assuming that a student didn't study hard enough or that they didn't understand the material, go deeper: were they having an off day on the day of the test? How were they feeling? Sometimes, the story can be more revealing than the grade or the test score.

Insight #50

In order to improve your grades and test scores, you first need to accept your current grades and test scores.

ONCE AGAIN, this probably sounds surprising coming from a math tutor and test preparation coach. After all, aren't you paying me to help your kids improve their grades and test scores? This is one of those situations in which, paradoxically, people often need to make a shift in the opposite direction of what they would think. In this case, the paradox is that in order to improve your grades and test scores, you must first accept your current grades and test scores. And not just on an intellectual level, but also on an emotional level. Because one of the main problems that I deal with is overthinking and overanalyzing grades and test scores.

Let's take a moment to just be. Accept yourself for who you are right now, including your (or your child's) current grades and test scores. If you have any resistance to this, take note of it, and become aware of it. Be gentle with it, and feel it without judgment.

This goes back to my parallel struggles with weight. For example, in order for me to lose weight, I had to first accept my weight for what it was at the time. Now once again, I know that goes against the basic principles of dieting, in that the goal of dieting is to lose weight. Paradoxically, in order for me to lose weight, I first had to accept my weight for what it was at the time. With that, I allowed myself to integrate nutrition, movement, and self-care to honor my body. From that acceptance, my weight naturally stabilized, albeit not to my original goal, but I still lost weight. Similarly, my friend who got engaged in 2017 had to accept being single before she could attract a healthy relationship. Before she met her fiancé in 2015, she was in a series of emotionally unavailable relationships that were toxic. What shifted was that by accepting herself for being single, she was able to attract her fiancé.

How does this apply to grades and test scores? Plenty. In order to make progress, you must first accept what is. Please note that just because you accept what is does not mean that you are not open to up-leveling. Because this is how I see improving your relationship with math. It is an up-level. You might or might not get the grades you expect, but the fact that you are shifting your relationship to math is an improvement in and of itself. If you continue to push for higher grades and test scores, you will continue to repel them and will make your relationship with math worse.

THE ROLE OF PARENTS AND TEACHERS

(Insights #51-61)

Insight #51

Math shame is a pattern in the family, not a problem in the child.

OFTEN WHEN I MEET with a parent or teacher about a student, they say something along the lines of this: "He/she has so much potential. He/she just needs to buckle down and study harder." And sometimes they add this: "It's such a shame. He/she is so smart." I know they're trying to be supportive and helpful, but notice that there's a great deal of shame within that statement, and not just because it uses the word "shame." It does not address the many mental and emotional subtleties that go into the process of learning math.

The truth is, it is rarely as simple as studying harder. Once again, in going back to my parallel struggles with weight, a friend once said that losing weight was as simple as eating less and exercising more. Technically, my friend was right, but so many mental and emotional layers go into a person's eating and exercise habits. It's the same principle for studying math. If you don't deal with the mental and emotional issues that are plaguing you, then you are rarely going to be able to study math most effectively. If a student is being told that they're not trying hard enough, then that can be a shame trigger, and can further perpetuate existing shame.

Among many people, there is a stigma attached to getting extra help or support. This doesn't just apply to math tutoring, or even to therapy or counseling. I remember when I went to Weight Watchers, I felt ashamed because I felt like a failure for not being able to simply "eat less and exercise more." Likewise, many students may feel ashamed for not being able to simply "buckle down and study harder." That said, we need to be aware of tutors and teachers who are shaming, and especially those who perpetuate the "just study harder" myth.

In many cases, traditional tutoring is a "Band-Aid" fix that perpetuates math shame, just like traditional dieting is a "Band-Aid" fix that perpetuates body shame. In other words, if someone goes on a diet, they might lose a few pounds at first (as did I), but ultimately, if they don't heal the underlying shame, they will regain the weight. Then, when they regain the weight, it becomes a self-fulfilling prophecy that reinforces the old shame story. Then they go on another diet, and the cycle starts again. Traditional tutoring and homework help can perpetuate a similar cycle with math shame. If the underlying shame, anxiety, and avoidance are not healed, then math will always be an emotional trigger. And the stakes will get higher as the math material becomes progressively more difficult.

Since shame is cumulative, the shame grows right alongside the difficulty of the material, so it becomes a double whammy. Eventually, both grow so overwhelming that the avoidance part begins to take over, and the person decides to only take the minimum amount of math required to graduate (usually some time in high school or college). Then they tell themselves the "I'm not a math person" story for the rest of their lives. It also goes without saying that punishing a child (i.e., grounding) or removing privileges (i.e., phone or computer use) as a consequence for poor grades or test scores also sends a shaming message, and should be avoided.

Insights #52-54

#52 - *The Math Lady also works with adults, not just kids and teens.*

#53 - *In order to combat math shame, we need to start with the parents and teachers.*

#54 - *When a parent or teacher heals their own math shame, it has a ripple effect on their children and students.*

MOST PEOPLE ASSUME THAT I work with kids and teens only. This is unfortunate because math shame affects adults of all ages. What's more, the adults (of all generations) are direct influencers of the younger generation. If these adults have unhealed math shame, then they will likely pass these negative feelings onto the next generation, albeit unconsciously.

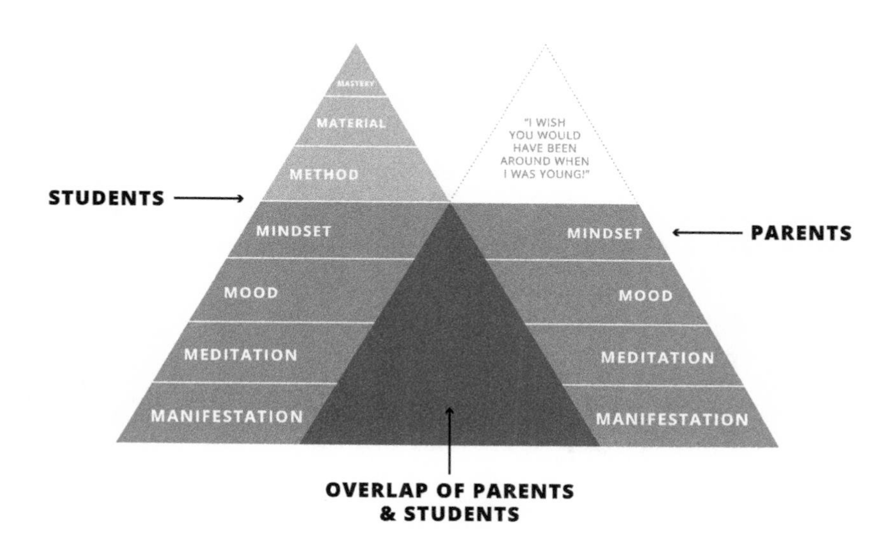

See the interconnected pyramids? It's our good old friend the 7M Pyramid, connected to another pyramid. Here's where the overlap between parents and children (or teens) becomes apparent. Obviously, when children and/or teens are studying math, they have a certain result that they want to achieve, and in order to do so, they need to master the material with an appropriate method. In other words, they need to study. With the parents, those first three levels are usually no longer relevant, so they're grayed out. But with the parents, they still have the lower levels. In other words, parents still have a mindset (thoughts) and mood (feelings) about math. And the spiritual levels (meditation and manifestation) can benefit both students and parents.

What's more, these levels can overlap between students and parents. In other words, a parent's mindset and mood about math can affect their children's and teenagers' mindsets and moods, and vice versa. That is why parents are so vitally important to breaking the cycle of math shame and math anxiety in the next generation.

When parents and teachers and other adults say to me, "I wish you would have been around when I was young!" it usually means one or both of two things: a.) They suffered in math when they were young, and b.) They believe that they were robbed of opportunities because of their negative experiences in math. These parents and teachers usually want their children and students to have the opportunities that were not available to them, and for them not to suffer as well. In order to increase achievement in children, we need to start by eliminating (or at least reducing) the suffering in children. Thus, in order to eliminate the suffering in children, we need to heal the suffering in their parents and teachers.

This is where the ripple effect comes in. You as a parent (or teacher) might not think that your past experiences in math have anything to do with your children's experiences in math today. But they are interconnected. The parents' past affects the students' present, and the students' present affects the world's future. In other words, think of the three spirits in Charles Dickens' *A Christmas Carol*. The past, present, and future are all intrinsically connected, and in order to change the future, we must start by reframing the past and taking action in the present.

I believe that it is vitally important to work with adults, in addition to children and adolescents. Not only because the adults have a direct effect on the younger generation, but also to help the adults realize that it's not too late for them. After all, most people think of math as an issue that is exclusively for children and students. That's where our culture has it wrong (or is not seeing the whole story). Math is for all ages and extends far beyond the classroom. STEM affects all of society, and vice versa, so in order to be a well-educated person (of any age), you need to have a basic understanding of STEM principles. Unfortunately, negative experiences in math often block both children and adults from appreciating its beauty. And these adults often think that it is too late for them. Although these adults are well-meaning, their limiting beliefs are often passed down to the next generation. In other words, the younger generation picks up on these beliefs and unconsciously absorbs them. That is why we have to start with the parents and teachers.

A prime case in point would be a middle-aged elementary school teacher whom I knew in 2007. When I was introduced to her, our mutual friend introduced us by saying that we were both teachers and that piqued her interest.

"Oh, really? What grade? First? Second? Third? I teach third, and the oldest I could handle is fourth because in some of those kids, the hormones start kicking in as early as fifth grade!" she said.

"Actually, I teach advanced calculus and physics to high school students. And one section of freshman algebra, but it's not my favorite," I replied.

"Goodness! That's surprising that you would choose to teach such advanced math and science!" she said.

"And why is that?" I asked.

"Because girls are usually stronger at arts and languages, and boys are usually stronger at math and science," she responded.

Before my jaw could drop to the floor, our mutual friend returned with the drinks and launched into a story of her own. They both seemed to have forgotten what had just happened. But I never forgot. Even eleven years later, that interaction has become a prime driving point in my career. Let's not forget that this woman was a third grade teacher. That meant that she was probably sending these sexist messages to impressionable 8- and 9-year-old girls. The very thought makes me shiver.

There is scientific evidence to back this up. In a 2010 study, which I repeatedly referenced in my master's thesis on gender and math anxiety, it was found that female elementary school teachers who had math anxiety themselves were much more likely to have female students who performed worse in math than those who had teachers who were more confident in math. In other words, the teacher's feelings rubbed off on the students, and the effects lasted for years. The same can be said for parents. In a 2015 study, it was found that parents who had math anxiety were likely to transmit their anxiety "like a virus" when they helped their children with math homework. This effect did not occur if they did not give homework help to their children.

These two studies emphasize the principle that parents' and teachers' attitudes and feelings about math directly affect their children and students, albeit sometimes unconsciously. That is why we need a greater awareness of adult math shame, and how it affects the next generation. On the upside, when a parent or teacher heals their own math shame, it has a ripple effect on their children and students, inspiring the next generation to heal their math shame. Or better yet, to prevent the development of math shame in the first place. After all, an ounce of prevention is worth more than a pound of cure. That is why it is best to start with the parents and when the kids are young. After all, the studies show that children can develop math anxiety as young as six years old.

That is why the Math Lady is proud to offer services to adults of all ages. After all, so many adults say, "I wish you would have been around when I was young!" (Implying that they hated math when they were younger, and still do today). And don't worry; I'm not going to torture you with calculus, algebra, geometry, or even fractions or times tables. Instead, we will reframe your past in math, and release it so that it no longer affects you or your children or students today.

I know this idea probably goes against everything you've been taught. You might even have some negative emotions triggered from this section, and from this book as a whole. In fact, I would be surprised if you didn't. Be gentle with yourself. Notice the feelings when they come up, and then try to reframe them, and see them in a new light. Most importantly, think of your children and/or students. Part of the reason you may have picked up this book is because you don't want them to suffer in math the way you did. That is why we need a new way that incorporates the mental, emotional, and spiritual aspects, and that starts with the adults in their lives setting a good example.

Insight #55

Parents play a vital role in breaking the cycle of math shame.

WHY IS THERE A section for parents in this book? I'm not the one who's studying. I'm not the one who's taking a test. I want something for my child or grandchild, not for me. Yes, I know that the most pressing problem right now is your child's struggle in math. You might have a story or two about your own experiences in math, but surely, they're not relevant to the current situation. Or are they? The past affects the present, sometimes in ways that we're not aware of. If you have painful experiences with math in your past, you might be tempted to brush them aside and never think of them ever again. For the sake of your kids, you might want to think again.

Math anxiety that starts in childhood can often persist into adulthood. As we said before, math anxiety is often rooted in math shame. Many adults carry that shame with them, decades after they've left the classroom. Whether they realize it or not, that shame can affect them in the present, and can even affect future generations. As an example, if you say that you're not a "math person," your children will hear that, and will likely interpret that as a green light to give up on math. Parents (and grandparents) can also perpetuate sexist beliefs about math ability.

For instance, to comfort their daughters who are struggling in math, many mothers say, "Don't worry, sweetie. I let your father (or stepfather) do the taxes." Now, let's say that the teenage daughter was struggling in English. Can you imagine the mother saying, "Don't worry, sweetie. I let your father (or stepfather) read me the *New York Times*"? No, because that would be ridiculous.

Speaking of taxes, for many women, anxiety about math in childhood becomes anxiety about money in adulthood. Not only that, but a lot of women (and men) don't realize their financial potential because they don't have the confidence to deal with the numbers that come with their finances. That lack of confidence can manifest as avoidance and procrastination when it comes to dealing with finances. Not that there's anything wrong with getting help for your taxes, or any other task. For example, I outsource my housecleaning and my car maintenance. The key difference here is that I outsource these tasks to free up my time, not to avoid anxiety. If you put off doing your taxes, I ask you to become aware of why. The answer might surprise you.

In addition to comments like, "I still use my iPhone (or calculator) to calculate a tip at a restaurant or bar," and "I just let my husband (or partner) do the taxes," I also hear a lot of women lament, "I always wanted to be a doctor or scientist (or nurse or psychologist or architect or any number of careers), but then the math got too hard, and I dropped out." Once again, think about the example that this sets for your daughters (or sons). Remember, as a parent, you set the tone.

Insight #56

Both students and parents can have resistance to math.

A COMMON REASON that many parents and teachers are picking up this book is because their students are resistant to doing work in math. It seems like without fail, they continue to make excuses and underperform. That is frustrating for you because you know that they're capable of doing the work, but they just can't seem to buckle down. So, what gives?

Paradoxically, pushing them will only make them shut down even more. They may have some inner resistance to doing math. What is resistance? It is a reluctance to doing something, and an unconscious payoff to staying stuck. We all have resistance to things. For example, as adults, we might be resistant to change or might be resistant to something as simple as going to the dentist or doing taxes. While this might look like procrastination, there is usually an inner psychological reason that is worth exploring.

Let's talk about a favorite topic: excuses. Excuses are frequently rooted in resistance. I once saw a social media post from a life coach, a former smoker whose father died of lung cancer. In this post, in no subtle terms, she called out smokers on their excuses for not quitting. Although technically she was right, she could have approached the topic a bit more tactfully, addressing the various factors that go into a person's resistance to quitting smoking. The same principle is true for resistance to doing math. What is the payoff to staying stuck in an unproductive pattern? Let's explore the difference between an explanation and an excuse. An explanation offers clarity, but provides a way for one to get "un-stuck," and to take responsibility. For example, in going back to the example of quitting smoking, some people might find a pattern of stress being a trigger for smoking. That's a wonderful example of an explanation, in that it gives the person awareness, and an opportunity to choose a better way. This example of stress can also be an excuse. In other words, an excuse keeps you stuck, and keeps you defensive and resistant, and keeps you in the cycle of evading and avoiding responsibility.

Excuses keep people stuck in their disempowering stories about math. This is true not only for students but also for parents and other adults. In fact, as ironic as it may sound, when parents try to push students out of their "excuses" about math, they might push them deeper into resistance. On the same token, this pushing might be a manifestation of the parents' own resistance to math.

This is yet another example of how parents and students are inextricably linked in the process of negative thoughts and feelings about math. Resistance can affect both students and parents, and it can mirror off of each other, creating a vicious cycle.

So once again, the first step is awareness. Become aware of the resistance. What does it feel like? A lot of times, resistance disguises itself as procrastination, or overwhelm. In other words, not knowing where to start. In that case, baby steps are your best friend. When you're starting with baby steps, temporarily ignore any deadlines. Just do one small thing to move yourself forward.

Insight #57

Telling your teen stories about how you hated math when you were young can backfire.

WHEN YOUR CHILD or teenager is struggling in math, it is a natural instinct for parents to sympathize, especially if they struggled in math when they were younger. A common way of bonding is for the parent to tell the child stories about how they were "bad at math" in school, but that they now wish they would have taken more math.

A common question that many teenagers (and even children as young as 10) ask their parents these days is, "did you smoke pot when you were a teenager?" This is especially true in light of some states legalizing marijuana in recent years. As we all know, many adults have experimented with marijuana and other drugs when they were younger, and many still imbibe today. Unfortunately, if a parent admits that they smoked pot when they were young, it often implicitly sends the message that it's OK for the teen to experiment with marijuana now. In other words, it is a reflection of the "do as I say, not as I do" phenomenon. The only way that this can be an effective tool with your teenager is if you admit that you smoked marijuana when you were young, and then reflect on how it affected you then, and why you've changed now. More importantly, it is best to emphasize how these inner changes in you can positively affect your teenager today.

It's the same phenomenon with math shame. If you have had painful experiences with math in the past, you might be tempted to share them with your child or teenager. But the thing is, if you haven't completely healed from them, your children can pick up on that, and can absorb your negative emotions surrounding these experiences. Before you share these stories with your child or teenager, it is vital that you do the inner work for healing and transformation. Otherwise, these stories just become a tool for perpetuating math shame throughout the generations. Although not as obvious as with the marijuana example, this is a classic case of "do as I say, not as I do." In other words, if a parent has not healed their math shame, then telling their child to buckle down and study harder is not going to mean anything. If anything, this mentality perpetuates math shame and might make the children resent their parents for holding them to a different standard than they hold themselves. In other words, parents need to lead by example, and math shame is no exception.

If you do want to tell your teenager stories about your own experiences in math, please do, but be careful of how you present them. In other words, before you share a story with your child or teen, do some reflection on it yourself. What did you learn from this story? How is this story affecting you now (if at all)? They always say that you know you have healed from a story when you can tell it without it triggering any negative emotions in you.

Insight #58

Labeling a child as an "A student" or a "B student" or a "C student" can make them feel pigeonholed.

WHEN ONE OF MY FORMER students was in ninth grade, she was adjusting to the new schedule and new teachers of high school, which were quite different from middle school. She was taking honors geometry from a woman who would later become one of her favorite teachers. At first, it was difficult for my student to understand her teacher's accent, and the questions on her tests were often ambiguously worded. Long story short, she ended up getting a B+ in the class for the first quarter, after she had consistently gotten A's in math in the past. After her father received the report card, the first thing he said to his daughter was, "you are not a B student in math," and he sent me a message on social media to request an appointment the following day. Upon talking to my student, she said that this interaction with her father made her feel trapped and pigeonholed, and intensified her pre-existing perfectionism.

On the other hand, putting a student in the box of "C student" can make them feel like they're not working hard enough, and can put them in the box of "underachievement," thus making them unconsciously perpetuate that pattern. My biology professor in college always said that it took a lot of work to get an A, and a lot of work to get an F, but not a lot of work to get a C. While he has a point, his theory does not apply to all students and all classes. Namely, a grade is a snapshot in time and one that can be easily changed. A grade is just a grade and does not define a person. Unfortunately, with all the weight that parents and teachers place on grades and test scores, this can be a difficult concept to internalize on an emotional level.

The point is, parents' and teachers' reactions to grades can emotionally affect students. These emotional effects can last a lifetime. In other words, it can become a self-fulfilling prophecy of sorts, in that we unconsciously believe the stories we tell ourselves. If you label a child as an "A student," they might beat themselves up if they get an occasional B. On the other hand, if you label a child as a "C student," he or she might not believe in himself or herself. These labels don't give students room to grow and change and evolve, which are key components to a healthy attitude towards math (or any other academic subject). Just as well, these labels perpetuate the "it's too late" limiting belief for adults. In fact, a friend once asked me if I told my current students (and their parents) my SAT scores. I said no because a test score from 1999 does not define me today as an adult, or even as a test preparation coach.

Insight #59

Students, parents, and teachers are all intrinsically connected in the process of math education.

I KNOW THAT SOME PARENTS and teachers might be annoyed with me now, and I applaud you for sticking with me. At this point, you might be fed up, and might still be skeptical about the role you play in your child's or student's math anxiety and math shame. After all, you might still think that this has nothing to do with you, and everything to do with your children and/or students.

There is a big difference between blame and responsibility. In today's culture of education, it is a culture of a triangle of blame, with students, parents, and teachers all affecting each other, and the politics of education affecting all three. In this triangle, parents blame teachers and students, teachers blame students and parents, and starting around middle school age, students blame teachers and parents. With this culture of blame, the lines of communication shut down, and the cycle of math anxiety and math shame perpetuates. In order to break this cycle, parents, students, and teachers need to work together and communicate.

First and foremost, children need to learn responsibility at a young age, ideally starting in elementary school, or even earlier. For example, I firmly believe in children having household chores, such as taking out the trash or feeding or walking the family pet. In addition, learning to manage money at a young age can help children to learn responsibility. More importantly, these lessons can be applied to their studies, whether in math or any other academic subject. You might have heard of the term "helicopter parent," with many parents doing things for their children well into their adolescence, and even their college years. This fosters a learned helplessness of sorts, and does nothing to help children and teenagers learn the concept of responsibility.

The responsibility falls not only on the children. Parents and teachers need to be role models of responsibility, and that includes being responsible for one's own thoughts, feelings, and reactions. Parents and teachers should work as a team, ideally starting in elementary school. When parents blame teachers, and teachers blame parents, then the students are stuck in the middle. As early as age six, they can pick up on this. When parents and teachers both blame the students, then the students can feel like they're being ganged up on. Healing math shame is a joint endeavor. It should have the support of all parties involved—students, parents, and teachers.

On a developmental note, if students are under the age of 12 (give or take), then the parents and teachers should be primarily responsible for setting the tone. That is why it is vitally important for parents and teachers to role model a healthy attitude towards math, and to become aware of their own feelings and reactions to math, in order to avoid transmitting them to their students and children. Because remember, students, teachers, and parents are all connected to each other in this process.

Insight # 60

Shifting from blame to responsibility is critical to breaking the cycle of math shame.

IN REFLECTING UPON some of my earlier insights, there are several themes that are important here. First, we need to define the difference between blame and responsibility. Blame involves holding a grudge and being in denial. On the other hand, responsibility involves taking accountability for one's actions and words, and changing for the better in the future.

When I first presented my theory of parents' math shame affecting children's math shame, not everyone was receptive. For example, a woman (who happened to be a mother) went on a tirade, accusing me of "blaming the parents." Ironically, she then went on to blame the teachers, and the school system as a whole, although I completely understand where she's coming from. Unfortunately, it is the culture of blame in our current education system that got us into this mess in the first place.

In going back to the concept of locus of control, blame focuses on an external locus of control. In other words, looking to outside people, institutions, and situations to point the finger at. Responsibility turns it around and focuses on an internal locus of control. In other words, what can we change? In the "Serenity Prayer" that is popular in 12-Step circles, it says, "grant me the serenity to accept the things I cannot change, the courage to change the things that I can, and the wisdom to know the difference." And basically, the things we cannot change have to do with other people, and the things we can change have to do with ourselves. In other words, shifting from blame to responsibility involves taking a stand, and stepping up to change the things that we can. Fortunately, whether you are a parent, a teacher, or a student, there is much more that you can change within yourself than you probably give yourself credit for. Even if you can't change a situation (like the educational system, for example), you can still change how you respond to it.

My high school English teacher once said that the word "responsible" is derived from "response" + "able." In other words, able to respond. This also resonates with what one of my spiritual mentors taught me in 2013, about the difference between responding and reacting. Sometimes, we have an urge to react immediately, but it is often wiser to wait 24 hours to reflect, and then respond appropriately. This way, your raw emotions don't cloud your response. Many students, as well as parents and teachers, can be very reactive when it comes to math because it is a shame trigger for them. Remember, we're all human. And part of being human is admitting to making mistakes, which is part of what responsibility is all about.

Insight #61

In order to encourage your students in math, it is best to avoid comparing them to others.

IN MANY AREAS OF LIFE, not just in math and academics, we compare ourselves to others. Even as adults, you might say to yourself, "she's thinner than me," or "he makes more money than me," or "the couple next door is taking a nicer vacation than we are." By comparing ourselves to others, we invite competition and insecurity about ourselves.

The same principle applies to math and academics. This can apply to students, parents, and teachers alike. For example, a parent might compare their two children to each other. An example of that would be, "your older sister got an A in algebra, so I can't understand why you can't, either." While statements such as these might be technically true, they invite the younger sister to feel inferior and to feel like there's something wrong with her in that she's not more like her sister. Similarly, students might also do this with each other, comparing themselves with their peer group. Just as well, parents might compare themselves to other parents, seeing their children as a reflection of themselves as a parent.

Although comparing oneself to others is a part of human nature, it is not a very healthy mindset. It invites feelings of anxiety, shame, and inadequacy, and triggers the AAS Triangle. We are all on our own timetables, and we all have our own strengths and challenges. Comparing oneself to one's sibling or friend is like comparing apples to oranges. You often don't know the whole story behind the other person. It was said that on social media, you see the highlight reel of someone else's life, whereas you see the behind-the-scenes of your own life. The same principle can apply to comparing grades and test scores. You don't know what's behind that grade or test score.

There is a fine line between seeing someone as a role model and comparing oneself to him or her. When seeing someone as a role model, it is important to draw the line between emulating and idolizing. As Eleanor Roosevelt always said, "no one can make you feel inferior without your consent." The truth is, comparing oneself to others just breeds negative feelings. The same is true for parents and teachers comparing their students to others (peers, siblings, etc.). In order to best encourage a student, it is best to see him or her as an individual, with a unique set of strengths to be nurtured.

What works for one person might or might not work for another. If your student is struggling in math, and a study method worked for their friend or sibling who got an A (or B) in that class, it might or might not work for the struggling student. It is always worth a try, but if it doesn't work, then don't force it. Ultimately, we are all on our own path, and the timetable that works for one person might not work for another.

COMMUNICATING ABOUT MATH

(Insights 62-67)

Insight #62

When a child feigns an illness to avoid taking a test, they are usually genuinely suffering.

"MO-OM! DA-AD! I'M SI-ICK!"

How many times have you heard that the night before or the morning of a big test? Those of you who are parents and/or teachers will probably know the drill. Your child/teen or student comes to you the morning of a big test (whether in math or any other subject they dread), and complains of a vague illness. You know better than to be manipulated, so you call them on their bluff, and instruct them to prepare for their test, because they're going to be taking it, ready or not.

While the child might not physically be sick with whatever they say they're sick with (a headache, stomachache, fever, menstrual cramps, etc.), it is important to know that faking an illness happens for a reason. That is, it is usually because they are psychologically suffering, and know no other way to express it. This suffering is often not acknowledged, and not only is it not acknowledged, but it is also invalidated, which makes things worse in the long run. If you're like most parents, you don't believe the kids, send them to school anyway, and tell them that it's not nice to lie about being sick. That's normal and is what most parents would do. After all, that's probably what your parents used to do when you were young. However, it is not the most effective way to approach this situation.

While feigning a physical illness might be an attempt at manipulation, it could also be because the child or teen is genuinely suffering. While they might not be suffering physically, they might very well be suffering emotionally. Sometimes, perhaps more often than we realize, the physical and emotional are inter-connected.

I have a confession to make. Although I have never had anxiety about math, I have had plenty of anxiety about social things when I was growing up. In other words, the way you felt about the SAT is probably very similar to the way I felt about the prom and other traditional social gatherings in high school.

Back in my high school days, I never feigned a headache or stomachache on the day of a math test, but I often did on the day of a dance or other social event. The thing is, I wasn't entirely making it up. Yes, I might have exaggerated the physical symptoms, but the emotional symptoms were there, and they were intense. So intense that I felt them physically. Just like many kids today are shamed for "faking sick" on the day of a math test, I was shamed for "faking sick" on the day of my high school Christmas dance. In both cases, that shame can potentially stick around for years to come.

Then again, you don't want to let them stay home. This is enabling behavior, and the stakes will get higher the older they get. So that is why it is important to address it early. This is where compassionate communication comes into play. Instead of shaming them for "faking sick," ask them about how they're really feeling, and open a dialogue. But also, gently explain that sometimes they're going to have to do things that make them uncomfortable (i.e., taking a test for them, or going to a social event for me). Even an awareness of this discomfort can go a long way.

Insight # 63

When communicating about math, I-messages are more effective and supportive than you-messages.

WHEN PARENTS AND teachers need to communicate with students about math, the ways in which they communicate can have a profound effect. This is also true for communication about virtually any topic. As we have discussed, math can be quite an emotional subject, and unless it is handled with care, further emotional collateral can occur, and can inadvertently keep both students and parents stuck in the AAS Triangle.

There are many communication techniques that can be employed in order to communicate about math in a manner that gets your message across but is also emotionally supportive. One of these techniques involves using what's called I-messages instead of you-messages. What are I-messages? What are you-messages? Put simply, I-messages contain the word "I," and focus on the speaker and how he or she is affected by the words and actions of the person being addressed. On the other hand, you-messages contain the word "you," and focus on generalizations about the person being addressed. The concept of I-messages versus you-messages can be used in virtually any other setting, not just in math, and can also apply to communicating with adults. For example, an I-message would be something like "I feel upset and hurt when I am not acknowledged on my birthday." By contrast, a you-message would be something like "You're so forgetful! Why do you always forget birthdays and anniversaries?"

Notice the tone of each of the above. As an example in math, an I-message would be "I am concerned about your study habits in math, because I worry about your future." Compare that to "You always wait until the last minute to start studying for a math test!" How does each statement feel to you? Note that while the first statement technically contains the word "you," the focus is shifted onto the parent's feelings, and how they are affected by the student's behavior.

You-messages often sound accusatory and are often blaming and shaming. You-messages often put the recipient on the defensive, and make the recipient shut down emotionally. Just as well, you-messages are often internalized as character flaws. Lastly, you-messages often (but not always) include extreme words such as "always" and "never," which we want to avoid in interpersonal communication. How does this apply to math? There are many times when a parent or a teacher might have to communicate with a student about their performance in a current math class, or their plans for a future math class. Here is where it is vitally important to use I-messages and active listening skills.

It can be challenging because I-messages are often more vulnerable than you-messages. After all, I-messages focus on the feelings of the speaker, and the speaker might or might not feel comfortable expressing his or her emotions. He or she might not be aware of his or her emotions and might find that the most direct way is through a you-message. Paradoxically, the direct route is not always the most effective.

Insight #64

Try asking "how?" questions instead of "why?" questions.

IN ORDER FOR YOUR communication about math to be most effective, we need to focus on the present and the future, and not the past. When asking questions, "why?" questions usually reference the past. For example, "why did you procrastinate on your studying?" In addition, "could" statements are usually more effective than "should" statements, for the same reason of shifting the focus from the past to the future. Ever heard the saying, "don't 'should' all over yourself?" Same principle with your kids and students in math.

As discussed in the previous sections, many parents and teachers carry math shame from their childhood into adulthood. Sometimes this shame is inadvertently transmitted to their children and students. Therefore, parents and teachers need to be careful when communicating with their children and students about math. Thankfully, there are some simple tweaks that you can use to make sure that your communication is as shame-free as possible. As a side note, these concepts can apply to communicating with your children or students about anything, not just math. Just as well, these concepts can apply to communicating with friends, co-workers, family members, and other adults.

In building on the difference between "I-messages" and "you-messages," know that "why?" questions are often you-messages in disguise. For example, "why did you wait until the last minute to start studying for your test?" Instead, try replacing "why?" questions with "how?" questions. For example, "how can I help you make a study plan for the next test?" Similarly, instead of asking, "why do you hate math?" try replacing it with, "how is this attitude serving you?" or "what is this attitude costing you?" (When I say "costs," I mean both literal and figurative.) Shifting away from "why?" questions helps to shift away from the past, and towards a proactive solution for the present and the future. Similarly, "should" statements often focus on the past. For example, "you should have studied harder" or "why did you play video games instead of studying?" The past is over. What's done is done. No use lecturing a person (including yourself) about what they should (or even could) have done differently.

A more effective approach would be to focus on the present and the future, instead of the past. As a completely non-math-related example, if your teenager walks in late, your first instinct might be to ask, "Why were you out so late?" or say, "You should have called or texted. Why didn't you call or text?" Notice all the you's, why's, and should's, and the focus on the past (even if it was five minutes ago, it was still in the past). To turn this around into a present and future-oriented discussion might look like developing a plan for the next time the teenager is running late. Just as well, the parent expressing his or her feelings of concern when the teenager is late and does not call or text.

The same principle applies to talking about studying math, or any other subject. Remember, we cannot change the past. We can only change the present and future. This applies whether the past event occurred five minutes ago or five years ago. For example, if a middle school student is failing math, it would be shaming to say, "but you always got A's in math in elementary school." In this hypothetical case, elementary school is in the past. The present is middle school, so focus on the present.

Insight #65

There is a big difference between intent and impact, but we also need to add a third part: insight.

THIS SECTION REVISITS the topic of communication, and how vitally important it is to the process of developing a healthy mindset and mood about math. After all, as we have seen before, students are greatly impacted by the words of parents and teachers, for better or worse. By listening to the stories of adults who have had bad experiences in math, these words can stick with them for a lifetime. Whoever said, "sticks and stones may break your bones, but words may never hurt you" is sorely mistaken. Words do matter, and this also applies to math. I prefer to think of it in terms of Maya Angelou's quote, "people will forget what you said, they will forget what you did, but they will never forget how you made them feel."

The difference between intent and impact cannot be overstated. In other words, intent is what is meant by a statement or question. On the other hand, impact is how the recipient interprets the statement or question that is directed to him or her. We all interpret things differently. In my life coaching class, we did an out-of-the-box coaching exercise in which we partnered up, and did a creative coaching method. One pair, which consisted of two mothers of teenagers, involved one woman putting on glasses and trying to walk around the room. She saw things blurry through the glasses, and it was meant as a metaphor to illustrate that her daughter sees things through different "lenses" than she does, and that she needs to be aware of her daughter's lenses. Similarly, her daughter also needs to be aware of her mother's lenses. In other words, we all have our own lenses through which we filter the world.

This is why it is vitally important to differentiate between intent and impact and to be aware that impact is unique for each individual, depending on their "lenses." This is where the third part, insight, comes in. That is, awareness of the other persons' lenses. As an example in math, if a parent tells a student that she needs to study harder, the student might interpret that as the parent criticizing her and giving her the message that she's not trying hard enough. She might feel like she is trying hard enough, and is frustrated that she can't seem to try harder, or that it exhausts or frustrates her. Then she might snap at the parent, saying that the parent treats her like a child.

In this case, insight is vital. Both parents and students need to be aware of how their words affect the other. Not only that, but also to have an awareness of why they react the way they do. For example, the parent might trigger something within the child, or vice versa. Similarly, if a teacher says something that reminds a student of what a parent or previous teacher has said to them in the past, it might also trigger a negative response.

Insight #66

Asking "are you sure?" is a surefire way to undermine your teen's confidence.

"ARE YOU SURE?" is a common question that adults ask children and teenagers. While it may seem harmless and well-meaning, it can take a toll on a child's autonomy and self-esteem. For example, my friend's daughter was excited about taking a whale watching field trip with her second-grade class. The teacher instructed them to wear warm clothes, so she wore the *Lady and the Tramp* sweatshirt that her grandmother had given her for Christmas. As soon as my friend pulled up to the school, her daughter's teacher stopped her and asked, "are you sure you're going to be warm enough?" My friend's daughter was then not allowed to go on the field trip, because she didn't have a "proper jacket."

This illustrates a common theme. That is, asking a child or teenager if they are sure is tantamount to doubting them. For example, "are you sure?" is a favorite phrase of one of my well-meaning friends. She asks this in reference to everything from cooking without a recipe to taking an alternate route to the airport. But whatever it was, I felt like I was being doubted and undermined. Inevitably, I would absorb that doubt as my own, and question my decisions. It is the same principle in math education. So many of my students over the years have been paralyzed by self-doubt, afraid to take even the first step in a problem when they weren't sure. I would ask them what they thought the first step was, and they would answer with trepidation, "I'm not sure."

Well, that's just the thing. You don't have to be 100 percent sure in order to be confident. Uncertainty is always going to be a part of life, and always going to be a part of math. There are many math problems where there's not one "correct" way, but several avenues to explore. And yet, our system penalizes children (and adults!) for uncertainty and keeps them paralyzed in a cycle of self-doubt.

Perfectionism is based on self-doubt. All too often, the "I'm not sure" students are afraid of making a mistake. Well, guess what? Mistakes are part of the learning process. Unless you get past your fear of making a mistake, you won't move forward. This was especially true in a program that I used, where the online tests had to be answered in order, and you couldn't move on to the next question unless you had answered the previous question. These students, paralyzed by uncertainty, would get stuck on the first question and would make no progress. I would tell them to pick an answer and move on because it's not the end of the world if you miss one question.

It's okay to be unsure. In fact, self-doubt is often worse than making one mistake (or ten mistakes or a hundred mistakes). When parents and teachers ask, "are you sure?" they are unconsciously perpetuating the cycle of self-doubt and perfectionism. Remember, it is okay to be unsure. Similarly, it is okay to say, "I don't know." What is not okay is to use that as an excuse to stay in the cycle of self-doubt and shame.

Insight #67

If your student is in the midst of a challenging problem, instead of asking "any luck?" try asking "any success?"

ALTHOUGH I RARELY had trouble in math when I was growing up, I would often get frustrated by non-academic things, such as putting together furniture, fixing electronics, learning to drive or ride a bike, or even straightening my hair. When I was frustrated by something and repeatedly trying to get it right, I was sometimes asked, "any luck?" That bothered me for several reasons. First, it reflected a lack of confidence in my abilities, which triggered my own lack of confidence in my abilities. Just as well, when I took a psychology class in college and learned about a concept called "locus of control," I realized that asking "any luck?" reflected the external locus of control. In other words, attributing events to an outside force, such as luck or chance. On the other hand, the internal locus of control attributes events to oneself, such as one's perseverance and strategy.

In applying this to math, or virtually any other endeavor, as one may expect, the internal locus of control is a lot healthier and promotes the growth mindset, as opposed to the external locus of control, which promotes the fixed mindset. Jack Canfield, author of *The Success Principles*, talks a lot about the concept of locus of control when discussing his 100 *Success Principles*. He even gives the example of being late to a meeting, and a person tending towards the external locus of control blaming traffic, the weather, or a car accident on the road. Alternatively, a person who tends towards the internal locus of control will take responsibility for not leaving sooner, and will choose to be happy while waiting in traffic.

In response to the question, "any luck?" I personally would replace it with "any success?" The term success implies that it comes from the person who is solving the problem, and not from an outside force. In other words, the term success implies the internal locus of control. Just as well, the external locus of control can be applied to blame, and the internal locus of control can be applied to responsibility. As we have said before, a shift from blame to responsibility can yield amazing results.

Once again, we can apply Jack Canfield's *Success Principles* to math. In other words, if we embrace the growth mindset and the internal locus of control, we can challenge ourselves to stretch beyond our comfort zone, and to be responsible for our own success. The concept of luck implies that an outside force is responsible for our successes (and failures), and is very disempowering. Remember, when you have a success, own it! Every success, no matter how small, is a victory. And every victory can be used to create a new, empowering story in math, and put another nail in the coffin of your old, disempowering story.

CREATIVITY, SPIRITUALITY, AND PHYSICAL ACTIVITY

(Insights 68-74)

Insight #68

Physical activity is crucial to increasing achievement and confidence in math.

ALL TOO OFTEN, adults have bad memories of high school gym class. Often as bad as high school math class. The point of gym class is to teach the importance of physical activity. This is very similar to recess in elementary school, which is being cut in some districts due to time and budget constraints. I am a firm believer in the connection between mind and body. While math might not seem to have to do with exercise on the surface, it is all connected. After all, in going back to the pyramid, we can see that emotional and spiritual well-being are crucial to success in math. Due to the mind-body connection, physical well-being is inextricably linked to emotional and spiritual well-being.

Taking a break to do something physical can work wonders in releasing frustration and anxiety. Unfortunately, with the way the modern classroom is structured, there is often not much time for physical activity, even at the elementary ages. Although ADHD is a legitimate condition that requires legitimate accommodations, many of today's students with ADHD have it worse than their parents and grandparents did because of the lack of physical activity. Physical activity stimulates areas of the brain that are responsible for critical thinking and creativity. In other words, physical activity takes you out of your head. This might seem a little paradoxical for a book about math. However, overthinking can be a cause for lots of problems in math, and lots of getting stuck and repeating the cycle of math anxiety and math shame. Exercise is a wonderful way to get out of that cycle and to get unstuck.

Remember the parts about the limbic system? Exercise calms the limbic system and reduces anxiety. I know a retired lawyer who loves to play golf because he says it gets him out of his head. Although he was always a very intellectual person, he realized that golfing helped him to put his brain on pause. While he could have been analyzing the angle of the ball and the club, he instead found that not overthinking was the best way to achieve success in golf. In other words, working with his body, and not against it.

Similarly, one of my other friends has two daughters who are very involved in ballet, and she sent me a video of a ballet dance that incorporated math. Like golf, ballet works best if you get out of your head, and into your body. Surprisingly, there are a lot of connections to math in both areas. Even just walking can calm the limbic system. That is why I recommend doing some form of physical activity before a major test, such as the SAT, ACT, or GRE. You can even study while you're exercising. Once again, exercise helps you get out of your head and helps you retain the material better, while you calm your limbic system.

Insight #69

Infusing art and music into math can make it seem less intimidating.

MANY PEOPLE THINK that math doesn't go with art and music. In other words, the old "left brain" vs "right brain" dichotomy. After all, many times in casual conversation, such as at a cocktail party, someone might say, "oh, I'm a writer. I'm not a math person," or they might say, "I'm an engineer. I don't do that right-brain artsy stuff." As for me, I am not just a mathematician, scientist, writer, and educator. I also cook, bake, sew, paint, knit, crochet, cross-stitch, and play the piano. A lot to keep me busy! Believe it or not, all of the above involve some degree of math, whether you use it consciously or not.

All of the above, as well as art and music in general, can be useful in relieving stress and anxiety. Remember the limbic system? Well, creativity has been shown to have a calming effect on this part of the brain. That can then clear the way for the prefrontal cortex to do its job with math and logic. Math can be very abstract, or it can be very visual. More often than not, the two go hand in hand. For example, algebra and trigonometry are usually more abstract, and geometry is usually more visual. Students who are strong in the arts can get lost in the shuffle when the math becomes too abstract. However, we can make the abstract concepts come to life in visual ways by connecting them to the arts.

For example, cooking and baking can be a good way to introduce fractions, as well as ratios and proportions. If you had to double a recipe, could you convert all the ingredients? What if the recipe called for 3/4 cup of sugar, and you needed to double or triple it? Presto. Multiplying fractions. Just as well, musical scales go in cycles of seven (A, B, C, D, E, F, G), and when you add the A of the next cycle, you've got an octave or a group of eight. Notice the root word "oct-" means eight, just like it does in the words octopus and octagon (you can point this out the next time you see a stop sign). Lastly, geometry is very useful in quilting and sewing. Not only shapes like triangles and rectangles, but also concepts like congruence and similarity.

In 2013, I had the privilege of hearing Temple Grandin speak at a conference in Florida, where she signed my copy of her book, *Thinking in Pictures.* For those of you who don't know, Temple Grandin is an animal scientist and livestock specialist and is perhaps the best-known adult on the autism spectrum. Although the autism spectrum manifests differently in different individuals, for Temple Grandin, it meant that she thinks almost exclusively in pictures. Indeed, that proved to be a great strength in her designs for equipment for herding cattle. On the other hand, her strongly visual mind posed a challenge for her in abstract math. Using her own personal experiences, as well as those of others on the autism spectrum who are strongly visual thinkers, she suggests that some middle school and high school students might be better off taking geometry before algebra. In other words, the opposite of the traditional sequence of courses. I admit, when I first heard this, I was appalled. After all, at the time, I had spent the better part of seven years lecturing students and parents about the importance of algebra skills. The more I thought about it, and the more I put it in the context of Dr. Grandin's books and life, the more it made sense.

Many kids (and adults) need math that is very concrete and visual, and easily get lost and distracted with math that is too abstract. By having a strong foundation in the visual aspects of math, they can more easily apply these concepts to the more abstract concepts that they might find challenging. Just as well, an integrated sequence of math can also be an option. In other words, covering basically the same content but integrating algebra with geometry, and vice versa.

Insight #70

The only way to learn a sport or a musical instrument is to cultivate a positive mindset, and then practice. Same principle with math.

MANY OF MY STUDENTS, both current and past, are involved in athletics, music, and the arts. For most of them, these activities are "fun," and math is "non-fun." The process of learning math is very similar to the process of learning a sport or a musical instrument. By applying that process to math, we can make a lot of shifts in our mindsets about math.

Nobody is born knowing how to play baseball or play the piano, or virtually any other skill that is acquired. It doesn't do any good to pressure yourself or to beat yourself up for not doing something perfectly. In sports and in music, mistakes during practice are to be expected and are treated as learning experiences. Unfortunately, even during math homework and problem sets, mistakes are discouraged. It is that kind of mindset that leads to pressure, perfectionism, and paralysis. Sports and musical instruments depend greatly on a positive mindset. If you make one mistake during practice, would you declare yourself to be permanently defective at the sport or musical instrument in question and give up forever? Probably not. Just as well, baby steps can be valuable. All too often, we pressure ourselves to make leaps and bounds. Often, one step forward is all it takes.

While mindset work is vitally important, so is practice. In other words, practice breeds familiarity, and familiarity leads to mastery. When it comes to memorization, I'm not a big fan, but the truth is, enough practice usually leads to facts becoming automatic in the brain. On the other hand, practicing with a faulty mindset can backfire. Just as well, you could make math like a game. Work on doing practice problems, and then reward yourself. Also, treat the practice problems as an experiment. All too often, we become attached to the outcome and freeze with overwhelm. All in all, just get started and have fun. What? Math, fun? Are you crazy? Believe it or not, it's all in the mindset.

Our views of math are so clouded by our current educational system that the fun aspect of math has been all but removed. If you keep that playful, experimental mindset, a lot can happen. And, practice needs to be consistent. It does not necessarily need to be daily, but let's go back to the sports analogy. What would happen if you only practiced before a big game? Or in the example of the piano, what would happen if you only practiced before a big recital? Chances are, your performance during the big game or recital would not be at its best. Not just because of your lack of familiarity with the material, but also in terms of your mindset and mood. Believe it or not, consistent practice helps to support all levels of the pyramid.

Insight #71

STEM (science, technology, engineering, and math) can be enhanced by turning it into STEAM (adding the arts).

UNLESS YOU'VE BEEN living under a rock, you've heard of the buzz surrounding STEM (science, technology, engineering, and math) fields, and why they're so important to tomorrow's career landscape. Former President Barack Obama and former presidential candidate Hillary Clinton have both extolled the benefits of STEM education as being essential for tomorrow's workforce.

We can take this a step further, and enhance STEM by adding the arts, thereby making it STEAM (science, technology, engineering, art, and math). As I have said before, I am a huge proponent of adding the arts to math and science education. In society today, there is a perpetual myth that science and art are mutually exclusive. This is rooted in the classic right brain versus left brain dichotomy, where the right brain is typically assigned to creative pursuits, and the left brain is typically associated with logic and intellectualism.

The reality is that it's not that clear-cut. Brain science has proven that both hemispheres of the brain interact with each other, and can enhance each other. This concept can also apply to math and the arts. Creative pursuits can not only help students (and parents!) understand math concepts, but math can also be used to enrich creative pursuits. In addition, in our test-based culture, creativity is sorely missing from math education. The way that math is taught is not very creative, and that perpetuates the myth that math is a "non-creative" subject. In going back through history, math has always been very creative and discovery-based.

How can we infuse creativity into the school-based system? Show how math can apply to the arts. For example, there is a great deal of geometry in art, such as Picasso's cubism. Geometry has many applications to architecture, such as the medieval cathedrals in Europe. In addition, in physics, when learning about light, students typically learn about the ROYGBIV spectrum, which makes a rainbow. This can then be applied to color theory in art classes and also to the color wheel. Fractals in nature have many roots in mathematics and can be seen in virtually everything, from the patterns in branches in trees, to the stars in the sky. Did you know that the patterns in fractals have been scientifically proven to reduce stress in the human brain? A pretty cool application of math! And there you thought that math was increasing your stress, not reducing it!

How can we apply this stress reduction to the classroom? Once again, in going back to STEAM (science, technology, engineering, art, and math), we can enhance it one step further, by adding relationships, making it STREAM (science, technology, relationships, engineering, arts, and math). This integrates not only creativity (the arts) but also integrates social and emotional learning (relationships). This goes back to the importance of stressing the vital role that parents and teachers play in breaking the cycle of math shame.

Insight #72

Art, music, and physical education (PE) classes can improve confidence and test scores in math.

IN BUILDING UPON the previous section, I firmly believe that creativity enhances skills in math, and vice versa. Unfortunately, we still live in a test-driven culture, and in that test-driven culture, art, music, and PE are often being reduced or even eliminated, for the sake of devoting more time and resources to test preparation in math and English. This is sad for several reasons. First and foremost, it promotes the toxic energy of the testing culture. Similarly, it puts too much emphasis on test scores and makes students feel as though they are defined by a test score. It affects parents as well, making them feel like their children's test scores are a reflection of them as parents. In addition, it promotes a mindset of scarcity and comparison.

While we cannot change the test-based education system overnight, we can change our perspective and approach. In other words, we can work with it, not against it. And most importantly, we can become aware of our own energy in this test-based approach. With that being said, it is a sad fact that test scores do matter, but not in the way you think. Paradoxically, adding more art, music, and PE can increase test scores, even though it seems like time is being taken away from math, English, and other core subjects. In order to increase performance, whether on a test or anything else, you need a balanced approach. Regardless of how we measure results, a balanced approach to education can help not only the whole person, but also individual subjects. Unfortunately, a lack of creativity and physical activity can stifle the flow of learning.

The addition of creativity and physical activity can help students achieve in math because these activities balance their brains and bodies. Remember, learning math is a holistic experience, and is affected by the mind-body connection. The arts and physical activity can enhance brain regions, and can help make math come more easily. Just as well, if a student is frustrated by a math problem, taking a break to do something creative or something physical can work wonders. Remember when I said not to study in the 24 hours before a major test? Doing something creative and/or physical can work wonders to get your brain ready for the big test.

Too much rigorous test prep can fry the brain. In other words, it activates the limbic system and increases anxiety. This can affect parents as well because parents and students energetically affect each other. So, if a student is in the middle of an intense test-prep session, their parents can absorb their energy. Just as well, if a parent is anxious on behalf of a student, then a student can pick up on that energy. Ditto for teachers. All in all, the arts and physical activity are vital to moving that energy, and to avoiding the limbic system overload that tends to happen with too much academic test prep.

Insight #73

Math is not just about memorizing formulas or doing calculations. It is about creative problem solving.

IN OUR SOCIETY, both among parents and students, there is a reputation of math as being not only non-creative, but also forcing students to memorize a large number of formulas. This then promotes cramming before a test, in an attempt to memorize the formulas. This then promotes anxiety, which causes forgetfulness. Then the whole cycle starts again, with the students frantically trying to memorize the formulas. Math is about so much more than formulas. Math is about discovery. When one discovers these formulas, they are much more likely to remember them. Even if you don't remember the formula outright, if you have discovered it, then you can likely reconstruct it by logical steps. These "formulas" are often simplifications of patterns that have been discovered by mathematicians in the past, and are ways of summarizing these patterns for future generations.

Our common test-based system, with its emphasis on memorization, takes away from the joy of discovery. Many parents that I have interviewed have said that a priority for them is that their children understand the material, rather than memorize it. This is a legitimate concern and one that I agree with. Upon a closer look, this dilemma of memorization versus understanding is still on the level of material in the 7M Pyramid. We still need to go deeper, to look at the method, and especially the mindset, mood, and spiritual levels.

In going back to the memorization versus understanding dilemma, studying with a goal of memorization promotes an unhealthy mindset and mood. This can include pressure, perfectionism, and beating oneself up for making mistakes. This then becomes a catch-22, in that the automatic negative thoughts trigger the negative feelings, and then this cycle takes over. Studying with a goal of understanding can promote a healthy mindset and mood. Not only understanding, but I would take it a step further and suggest a goal of discovery. This will ultimately promote a growth mindset.

Another myth is that math is just about calculations. Once again, nothing could be further from the truth. In real life, mathematicians and scientists use calculations to support discoveries and patterns, not the other way around. In other words, math is used as a tool, and not as a crutch. In fact, most mathematicians, scientists, and computer programmers use calculators and computers to do their tedious calculations. The difference lies in your intention and mindset in using the calculator. In other words, you have to be careful about whether the calculator is a tool or a crutch.

Are you using the calculator to support your discovery, or are you using it to cover your negative feelings about math? Getting stuck in the calculator cycle can also promote an unhealthy mindset and mood, including frustration and paralysis. In this case, it helps to take a step back, and look at the whole picture: what problem are you trying to solve?

Insight #74

Humor can be a great antidote to math anxiety.

MATH HUMOR CAN GO a long way toward mitigating math anxiety. Remember our old friend the limbic system? Well, humor suppresses the area of the brain that is responsible for activating anxiety and shame. That said, humor can be a great tool for alleviating this anxiety and shame, at least temporarily. You might think of math as an "un-funny" subject, but we in the math world have lots of jokes based on wordplay and puns. Ditto for chemists and physicists. Ever heard of the "chemistry cat" online meme? He can help you study for your chemistry test, and alleviate your chemistry anxiety. (Remember, chemistry anxiety and physics anxiety are just math anxiety's cousins).

Every year, March 14th is Pi Day. For those of you who are uninitiated, one of the most fundamental constants in mathematics is the irrational number pi, which is approximately 3.14. Hence, celebrating it on March 14 (3/14). Pi Day is a great opportunity not only for education about the history, background, and importance of the number pi, but also a great opportunity for humor and clever jokes and puns. For example, pi is an irrational number (it can't be written as a fraction of two whole numbers), so there was a joke about pi being irrational. On a more serious note, math anxiety is a lot like pi: irrational, but very important.

Remember the AAS Triangle? And APR? Those are both plays on words with math acronyms. While the Math Lady's AAS Triangle has nothing to do with sides, angles, sides, or cosines, and the Math Lady's APR has nothing to do with credit cards and interest rates, they can be good mnemonic devices for the math definitions. Just as well, remember the limbic system? Humor activates the limbic system, so you can't be laughing and anxious at the same time. In the meantime, get creative. Anything you can laugh at is good progress in alleviating anxiety!

As a word of caution, it goes without saying that I would avoid humor that is based on the "bad at math" stereotype, and especially generalizations about women being "bad at math," as well as generalizations about racial and ethnic groups. This perpetuates the fixed mindset and does more harm than good, even if it is well-meaning. This can undermine women's self-confidence, and perpetuate sexist stereotypes. If we are to grow in eradicating math anxiety and math shame, sexism has absolutely no place, and neither do racism and homophobia.

A SOCIAL AND POLITICAL CONTEXT

(Insights 75-88)

Insight #75

In order to make progress in eradicating math anxiety, we need to challenge sexist stereotypes.

AT A HOLIDAY PARTY IN 2016, shortly after the presidential election, a guest proposed an interesting game. A conservative man in his mid-sixties, he started launching into a discussion about Hillary Clinton and Donald Trump. He then suggested that we go around the table, and say whether or not we think a woman should be president in the future, and why or why not. Then he added a twist: if you're a man, you're going to say "yay," and if you're a woman, you're going to say "nay."

When it came to my turn, I refused to answer his question and stated that the fact that it had to be asked in the first place was a reflection of our society's sexism. He then asked, "are you yay or nay? You were supposed to be nay." I then responded that I was neither "yay" or "nay," and that I refused to answer the question on the grounds that it promoted sexism. He then mumbled, "you're missing the point," and the conversation turned to another topic. Ironically, I felt that he was the one who was missing the point. That is, gender should not be a factor in selecting the president of the United States. Similarly, gender should not be a factor in math and science classes either. Unfortunately, it often still is.

As upset as this particular incident made me feel, it got me thinking. Namely, about the parallels between sexism in politics and sexism in math and science. It goes without saying that sexism was a major factor in Hillary Clinton's loss to Donald Trump in the 2016 presidential election. After all, the "Make America Great Again" rhetoric included a great deal of misogyny, not the least of which was referring to Ms. Clinton as the B-word. In political positions, women are judged much more harshly than are men, and are held to much higher standards. Just as well, if an over-qualified woman and an under-qualified man are competing for the same position, it often goes to the man.

The same is true in math and science fields. This is true across the board, whether in college classes, in admissions to graduate programs, or on the job. This usually starts in the home, with parents inadvertently perpetuating sexist stereotypes about mathematical and scientific abilities. Similarly, many teachers also perpetuate these sexist stereotypes, even as early as first grade. What's worse is that the perpetrators of these stereotypes are often women themselves. For example, two women at the party actually played along with the game and came up with arguments about PMS, menopause, and pregnancy, as well as the stereotype that women tend to be more emotional than men. These stereotypes can hurt women across the board, not only in politics but also in math and science.

Insight #76

Patriarchal values and assumptions undermine women in STEM.

WHEN I ORIGINALLY POSTED this insight on Instagram, a young woman reposted it with a caption telling her story of being a medical student, and people at the hospital assuming that she was a nurse simply because she was a woman. This was a great compliment to me because her story resonates with so many women. And she is not alone.

In 2017, a fifth-grade classroom had an assignment on vocabulary words with the "-ur" sound, and one of the blanks to be filled in was "hospital lady." Since the word in question had to have the "-ur" sound, one student filled in the blank with "surgeon." Unfortunately, that student was marked down, because the "correct" answer according to the answer key was "nurse." This assignment then made the rounds on social media, with many outraged women (and men!) commenting and sharing.

Both of these incidents illustrate the concept of patriarchy or the assumption that men hold the majority of power in society. Patriarchy has been a part of most cultures for centuries, and to this day, it affects women not only in terms of relationships but also in terms of their careers and finances. This is especially true for women in STEM fields, which have traditionally been male-dominated. Although strides have been made in recent years to get more women into STEM, the fact remains that in most STEM fields, the majority of professionals are men. This is especially true in computer science and engineering.

Even in 2018, assumptions still exist that boys and girls are wired differently and that due to these differences in brain wiring, boys are naturally better at math. This has been disproven by many scientific studies, but the myths still persist. If one boy's brain has been studied, it does not represent all boys' brains. Ditto for girls' brains. This illustrates the principle that correlation does not imply causation. Unfortunately, these assumptions still persist. One time when I gave a speech about sexism and math anxiety, I went into great detail about these stereotypes during my speech. During the Q and A session, a man in his mid-sixties asked the question, "Are girls more likely to have math anxiety because they have less aptitude in math than boys?" This question was asked in spite of the fact that the main point of my speech was that the gender socialization and myths of the patriarchy are the primary contributors to math anxiety in girls. I also had a similar experience with an Uber driver around the same age, who questioned my assertion that math anxiety in girls is due to social conditioning, and not biological differences. As I tried to explain, he kept interrupting me. At that point, I asked, "how many minutes to the airport?" and then put on my headphones.

Both of these incidents are understandable, because both of these men grew up in a different era when these gender assumptions were taken as fact. Now we know that scientific studies have proven that male brains and female brains are not all that different. In spite of this, there still exists a difference between men's and women's achievement in math, and that is largely due to patriarchal values and assumptions.

Insight #77

Benevolent sexism is still sexism.

NOT ALL SEXISM IS OUTRIGHT discrimination or insults. In fact, some sexism can be disguised as kindness. For example, if a man opens a door for a woman (but not another man). Similarly, if a man offers to help a woman (but again, not another man) with lifting heavy boxes or setting up technological devices. Lastly, if a man offers to drive when the passenger is a woman (but again, not another man). Many men think that the above actions are simply manners or chivalry, but upon a deeper look, they are rooted in sexist beliefs.

The above-described behaviors, among many others, are known as benevolent sexism. In other words, sexism that is disguised as kindness, but is really men subtly patronizing women. These behaviors promote the stereotype that women are the weaker sex, and need extra protection from men. Just as well, women (especially mothers and grandmothers) can also perpetuate benevolent sexism under the guise of being helpful.

For example, if a mother discourages a daughter from pursuing math and science because it's going to be "hard," she is subtly reflecting her lack of confidence in her daughter. Remember, "hard" is just a mindset; a mindset that can be shifted. Just as well, the stereotype that women are more "nurturing" and "compassionate" than men hurts women in the workplace and also hurts men. This is especially evident in parental leave policies, which heavily favor the mothers. If a father (whether gay or straight) wishes to take family leave, he will sometimes have a more challenging time having his request granted.

Benevolent sexism is a lot more subtle than hostile sexism. An example of hostile sexism would be Donald Trump's inappropriate comments about doing something to a part of a woman's body that I won't mention. An example of benevolent sexism would be Mike Pence's policy not to dine with a woman other than his wife. In the latter case, it is under the guise of protecting women. However, it still reinforces the belief that women need to be protected.

Even in STEM fields, women aren't immune to benevolent sexism. For example, the scientist Yvonne Brill's obituary began with "she made a mean beef stroganoff" and went on to describe how she cared for her husband and children. Only at the end did they mention her scientific achievements. There was nothing wrong with mentioning her culinary abilities or her devotion to her family, but when those aspects of her life overshadowed her scientific accomplishments, then that was a reflection of inherent sexism. To illustrate this point, a parody of Einstein's obituary was created, in which his family was mentioned before his scientific achievements.

A lot of traditional gender roles in heterosexual relationships are based on benevolent sexism. This puts men and women in boxes and excludes LGBTQ couples. Even in STEM fields, women are often assigned tasks such as taking notes, bringing lunch, or making the Starbucks run. Once again, there's nothing wrong with these tasks, but when these tasks are distributed based on gender, that's when it veers into benevolent sexism.

Insight #78

Sexism in other areas (religion, politics, family life, etc.) can have a ripple effect into sexism in STEM.

SEXISM PERMEATES OUR society as a whole. Consequently, sexism can affect the world of STEM. You might think that sexism in other areas has nothing to do with sexism in STEM. It is important to make the distinction. The basic principles of sexism are pretty consistent throughout, and all areas affect each other. In other words, sexism in politics can affect sexism in STEM, and vice versa. Ditto for sexism in family life and religion.

A prime example is the 2016 presidential election. There is no denying that sexism played a major role in the loss of Hillary Clinton to Donald Trump. Not only that, but sexism was a major theme throughout the election season, most notably in the debates. On several occasions, Mr. Trump rudely interrupted Ms. Clinton as she tried to make an articulate point. As he rambled on, she would look at him, and patiently wait for her turn to speak. In a way, Ms. Clinton's experience during the debates is similar to many women's experiences in the workplace, and even in their home lives. In other words, a woman is trying to make a point, and a man rudely interrupts her. Then the woman feels that she has to wait patiently for the man to finish before she can respond. Even when she does respond, her words are criticized much more harshly than those of the man.

This dynamic also happens in the fields of math and science, both in educational and professional environments. Often, during meetings, the men get all the speaking time, and the women have to patiently wait their turn. It's not right, but it definitely exists, both in politics and science. In other words, women have to work twice as hard to prove themselves in both fields. In both fields, when an overqualified woman and an underqualified man are competing for the same position or honor, it often goes to the man, in spite of the woman's qualifications. So, what gives? Once again, this is rooted in the patriarchy.

This behavior is usually learned in the family home, in which daughters learn from their mothers that it's OK for their fathers and stepfathers to interrupt them. Further, sexist assumptions about roles in the family can have a negative effect on women in the workplace, especially in math and science. More specifically, even in STEM fields, which are traditionally considered masculine, women are still expected to be the primary caregivers for their offspring. This hurts not only mothers but also fathers. If a new father wants to stay home with his child while his female partner goes back to work, many people give judgmental looks, and even make it difficult for the man to get time off from work. Similarly, gay and lesbian couples might face similar dilemmas.

In going back to the theme of sexism in STEM, sexism in all areas is connected, because it is all rooted in the patriarchy. In order to dismantle sexism in STEM, it is vital to take a look at sexism in society in general, and to dismantle those sexist attitudes in families and the political climate as well.

Insight #79

Sexist stereotypes in STEM frequently begin during childhood, and go unchallenged by parents and teachers.

AS WE HAVE SAID BEFORE, math shame is a problem that almost universally starts during childhood, whether we're aware of it or not. Many of those cases of math shame are rooted in sexist stereotypes about math ability. Similarly, those stereotypes also begin at an early age, and the earlier that girls are exposed, the more likely they are to absorb these negative stereotypes.

A good example of that is the way that women in STEM are portrayed in the media. For example, the popular television show *The Big Bang Theory* has two female characters who are very different. Amy is a stereotypical science nerd, wearing dowdy clothes and glasses, and acting socially awkward on several occasions. Her boyfriend Sheldon is equally socially awkward, and the two of them together perpetuate many stereotypes about math and science professionals. Penny is a model and actress who is blonde, fashionable, and not very bright. In other words, the stereotypes of the smart one and the pretty one. Even going back to the 90's, on the TV show *Step by Step*, the two sisters Dana and Karen embodied the stereotypes of the smart one and the pretty one, respectively. Indeed, there was a lot of sibling rivalry between the two, with Dana jealous of Karen's beauty, and Karen jealous of Dana's brains.

In addition, the toys that girls play with often have hidden stereotypes in them. For example, many little girls (and some little boys) have played with Barbie dolls. A few years ago, there was a talking Barbie doll that said that math was hard. Unfortunately, a lot of little girls absorb these messages, and take them as truth. In fact, one time when my friend posted a picture of her 10-year-old daughter, deep in thought in her fifth-grade math homework, a mutual friend commented, "math is hard." It is often taken for granted among women that math is supposed to be hard, and that limits their opportunities in math and science. In addition, remember the study of female elementary teachers and their math anxiety? Their female students, even as young as six, can pick up on their ambivalence. The younger this starts, the less likely they are to question them, and the more likely they are to internalize them.

What do we do? If you are a parent or a teacher, actively challenge these stereotypes, and talk to your daughters about them. For example, if they are watching a TV show or a movie that portrays the stereotypes of women in math and science, talk to them about it, and ask their opinions on it. Also, emphasize that math often has a bad reputation, but it doesn't have to be that way. In other words, challenge the stereotype that math is supposed to be hard for girls, and find ways to make it fun and easy.

Insight #80

Making pink STEM toys specifically designed for girls can reinforce sexist stereotypes.

ON THE SURFACE, it might seem like a good idea to have pink STEM toys specifically designed for girls, even as young as preschool. After all, we want to promote gender equality in math and science, and we want to get girls interested in math and science. The younger we start, the better.

Be careful what you wish for.

These pink versions are often not only watered-down versions of the "boy" versions of the same toy, but the colors and packaging in and of themselves promote sexist stereotypes. For one thing, the stereotype that pink is for girls and blue is for boys, which can be seen as early as the newborns in the hospital, or even in the trend of "gender reveal" parties during pregnancy, in which a cake is cut to reveal pink or blue frosting, or a box is opened to reveal pink or blue balloons.

In addition to being a mathematician, scientist, educator, and writer, I also knit and crochet. And since I'm at the age when a lot of my friends are having babies, I knit and crochet a lot of baby blankets. I used to automatically do pink for girls and blue for boys, but now I ask if there's a theme for the nursery. For example, one couple had a jungle theme, so I made theirs green and orange.

In going back to STEM toys, they should be gender neutral, including colors and packaging. When STEM toys are designed to relate to stereotypically feminine interests such as cooking and sewing, it sends the message that girls are supposed to be interested in these pursuits, and boys are not. These pink, watered-down toys also hurt boys, in that they penalize boys who are interested in so-called feminine pursuits. How often is a boy teased and assumed to be gay because he is interested in fashion, knitting, sewing, or cooking? In the American Girl store, I noticed that they had dolls for cooking, sewing, dancing, music, and parties, but no dolls that represented math and science careers. This was disheartening for me, because many young girls shop at the American Girl store, and see the dolls as role models. If there are no role models in math and science, then how can they grow up to believe that they can do math and science?

Lumping boys and girls into stereotypes can limit opportunities for both boys and girls. It especially hurts girls because girls have been historically disenfranchised. Gender roles are reinforced at a young age, albeit unconsciously. I know that some educators are of the camp that girls and boys learn differently. This is a generalization, and one thing I learned in my psychology classes is that correlation does not imply causation. Thus, if some girls tend to learn a certain way, it does not necessarily apply to all girls. Ditto for boys. Making toys and learning guides based on stereotypes only hurts both girls and boys and lumps them into categories.

Insight #81

LGBTQ youth are less likely to pursue STEM fields than their heterosexual peers.

JUST LIKE WITH THE SECTIONS on other controversial topics, I must begin with a disclaimer. I am aware that there are varying viewpoints when it comes to the subject of homosexuality. That said, if you are morally opposed to homosexuality for religious reasons, or are politically opposed to same-sex marriage, I understand your viewpoint. My point is not to debate the political or moral aspects of homosexuality. It is to shed light on a population whose needs are often not met, especially in the world of STEM.

There is a major lack of inclusivity of LGBTQ youth not only in STEM fields but also in high schools and colleges in general. For example, LGBTQ youth do not often have role models in STEM fields, and often have limited visibility in these fields. Unfortunately, Sally Ride, the American astronaut who passed away in 2012, was the exception and not the rule. Throughout her career as a professional astronaut, she often had to keep her relationship with her longtime female partner under wraps, as homosexuality was not well accepted at NASA back in the 1980s and 1990s.

Although much has changed since then, including the legalization of same-sex marriage and more public acceptance of LGBTQ relationships, we still have a long way to go. Namely, many parents and teachers assume that everyone is heterosexual, and do not acknowledge the possibility that some young people may be gay or lesbian. In addition, this issue was tackled in a 2015 article titled, "Why Is Science So Straight?" LGBTQ youth are far more likely to be bullied than their heterosexual peers, and many bullies directly target their sexuality. This bullying can reach a point of becoming unsafe, so much that the teen skips school. If they're being bullied, how can they concentrate on their studies, especially in math and science? Thus, it makes sense that LGBTQ youth, especially those who are being bullied, may be more likely to fall behind in their studies.

Similarly, in the adult workplace, heteronormativity exists as well. That is, assuming that everyone is heterosexual, without acknowledging the possibility that somebody might be gay or lesbian. For example, if a woman is wearing a wedding band, she might be asked, "so what does your husband do?" Or if a man is wearing a wedding band, he might be asked, "so what does your wife do?" If the person in question is in a same-sex relationship, such an assumption could put them on the spot and could make them feel unsafe. Due to the logical nature of math and science, many STEM workplaces discourage discussion of one's personal life, and that lack of discussion may invite assumptions, including the assumption that everyone is heterosexual. Thus, gay and lesbian people in science might feel unsafe outing themselves and might feel the need to stay in the closet.

We can alleviate this by bringing an awareness to LGBTQ people in science. This is especially important for LGBTQ teenagers, who need role models who are like them. Just like women in STEM need female role models, LGBTQ youth in STEM need LGBTQ role models. We can support LGBTQ youth by making sure that classrooms are inclusive. This includes math and science classrooms.

Insight #82

Mental health care is essential to keeping our most talented individuals in STEM.

TO START WITH A disclaimer, I am not a psychologist or a psychiatrist. That said, I am not able to diagnose or treat any mental health disorders. If you suspect that yourself or a friend or relative is suffering from a mental health disorder, please get professional help as soon as possible (and tell a parent if you're under 18). If you or a loved one are having suicidal thoughts, please put down this book and get help immediately.

With that out of the way, it is critical to address the topic of mental health in STEM fields, especially in women. Although it's stigmatized, it happens more frequently than we'd like to admit, among both men and women. This is an important part of why we need to break the stigma of mental health issues in STEM fields. Because STEM is so rational and logical and linear, emotions are often seen as a weakness. Thus, mental health problems are seen as a weakness. This is true not just in STEM fields, but also in society in general. This is a shame, because mental health deserves the same attention as physical health.

Depression, anxiety, and other mental health issues can be thieves of STEM careers. People, especially women, are often looked down upon as weak if they reveal that they have mental health issues. Even seeing a therapist carries a stigma, especially in the math and science world.

A number of women (and men) drop out of STEM majors and careers because of mental health issues. To make matters worse, these mental health issues can be self-fulfilling prophecies, just like the "bad at math" story that so many women tell themselves. This is all related to the cycle of anxiety, avoidance, and shame. For example, depression can cause many people to stay stuck in the state of avoidance. Then it becomes a purgatory of sorts and is difficult to get oneself out of. But it can be done. That is why it is vitally important to have access to mental health care, and to erase the stigma of mental health concerns, especially as they relate to women in STEM.

How do we know the difference between typical math anxiety and a genuine mental health disorder? After all, jitters about a math test are normal, and should not be taken lightly. When it affects other areas of your life, you should probably consider getting professional help. All in all, breaking the stigma of mental health care will help and support women (and men) in STEM fields.

Insight #83

Access to reproductive health care greatly increases women's opportunities in STEM.

BEFORE I BEGIN THIS INSIGHT, I am going to address the fact that the issue of reproductive health care is a controversial issue in America. I am aware that there are varying viewpoints, and that we should respect these viewpoints, and discuss them civilly. Since this is a book about women's opportunities in STEM, I feel that it would do an injustice to women, and especially young women, if this issue was not mentioned. That is, when women have the rights to control their reproductive futures, then all of society benefits. This is especially true in STEM careers.

Throughout the years, I have seen many promising young women go into STEM fields, only to drop out due to an unexpected pregnancy. What's even sadder is that these pregnancies could have been prevented with widespread access to birth control, and affordable options for women of all ages and economic backgrounds. Not all women in America (or in the rest of the world) have access to all reproductive choices, and when this access is denied, not only do these women and their families suffer, but their future careers suffer as well. In turn, the entire world of STEM suffers, and the world potentially misses out on life-changing discoveries.

In its most basic sense, controlling women's access to birth control really boils down to controlling women's behavior. If you think about it, this is rooted in the patriarchy. In turn, this limits women's opportunities and limits women's participation in STEM. When women's opportunities are limited, they are then confined to the traditional roles that the patriarchy prescribes for them, and the patriarchal cycle continues.

A lot of this is an economic issue. Because many women and their families cannot afford reliable birth control, they have limited options when an unexpected pregnancy occurs. Then when they are faced with an unplanned pregnancy, they are less likely to pursue STEM careers. Thus, they stay in the cycle of poverty.

Obviously, the answer to this problem would be to expand access to reliable birth control. Unfortunately, in the U.S., we are currently dealing with a political climate that is hostile to access to birth control. This is especially true in many so-called "red" states with limited access to family planning clinics. That is why it is important to continue to protect reproductive rights on a social and political level.

In the meantime, we can empower women to make choices for themselves and can continue lobbying for access to birth control, as well as health care in general. In addition to reproductive health care and mental health care, general health care is also important to keeping women in STEM. This is why we need affordable access to quality health care for all people, including affordable access to reproductive health care.

Insight #84

White, heterosexual men in STEM would do well to acknowledge their privilege.

THERE'S NOTHING WRONG with being white. There's nothing wrong with being straight. There's nothing wrong with being male. We need to recognize that in our society, and especially in math and science fields, that these are privileged positions. It would do everyone good to recognize this privilege and to not abuse it. I am well aware that the term "privilege" is a hot-button, emotionally-loaded, politically-charged buzzword. It would behoove us to examine the concept of privilege, and how it applies to opportunities in STEM fields.

Privilege is any advantage that is gained by means other than merit, usually through an inborn characteristic such as race, gender, or sexual orientation. With the way society has been structured for centuries, certain groups have been given implicit privileges simply on the basis of these characteristics. For example, in our patriarchal society, men are generally more privileged than women. Similarly, white people are generally more privileged than African-Americans or Hispanics in the United States, even in 2018. Just as well, gays and lesbians are often disenfranchised and do not enjoy the same privileges as heterosexual people. This is true even with same-sex marriage being legal in all 50 states, plus many other countries.

What does this have to do with math and science? Plenty. These societal privileges trickle down to the worlds of math and science and are perpetuated in the culture of STEM college majors and workplaces. For example, as this book goes to press, it has been less than a year since the scandals with Harvey Weinstein and his sexual harassment and sexual assault cases have leaked to the public. Unfortunately, there are all too many Harvey Weinsteins in the world of math and science, and many women in STEM drop out due to feeling unsafe because of sexual harassment.

There is no such thing as reverse racism, reverse sexism, or reverse homophobia. That is due to the very concept of privilege. In other words, the privileged positions (men, whites, and heterosexuals) cannot be discriminated against, because they hold the power. Many times, people complain that affirmative action programs discriminate against whites. Ditto for women's programs discriminating against men. This is not the case, because in general, the whites and males have the power. These programs are meant to help the non-privileged groups take back their power, and have equal opportunities.

Being in one or more of the above non-privileged groups can exacerbate math anxiety and math shame that already exists. Because so much of math achievement is measured by tests, we need to be careful that these tests are not biased against women and/or minorities. If they are, it becomes a catch-22 of sorts. People from white families usually (but not always) tend to have more economic resources than those from African-American and Hispanic families, and that can also skew results. That said, if you are a white man in math and science, I invite you to acknowledge your privilege, and to use it for good.

Insight #85

Discouraging girls from going into STEM by saying "you'll scare the boys away!" is disempowering for girls and is not inclusive of LGBTQ youth.

YOU MIGHT HAVE seen the above statement and assumed it was a relic from the 1950s. Or perhaps the 1980s at the latest. Unfortunately, many people often say something to that effect, even in jest. Even if they don't say it, they sometimes imply it.

This statement not only reflects sexist and heteronormative stereotypes but also affects impressionable young girls. It reinforces the limiting belief that boys don't want a girl who's smarter than them, and leads girls to downplay their intelligence, especially in math and science. This especially occurs during middle school, when social pressures intensify exponentially. And with the added stress of puberty and hormones, gender roles and expectations often intensify during middle school. This statement also assumes that all girls are interested in boys, which is not true. If a girl knows that she is a lesbian, or is even in the questioning phase, she might feel less safe coming out if her parents say (or imply) things like that. It sets up the dynamics of traditional gender roles in heterosexual relationships and excludes not only gays and lesbians, but also those heterosexuals who do not comply with gender roles and stereotypes. With the social pressures mentioned above, it is especially risky for LGBTQ youth (both girls and boys) to come out during middle school, and high school isn't much easier.

At this point, you might be asking about single-sex education. I went to an all-girls high school and an all-women's college (even though we shared a campus with several other co-ed colleges). It worked wonders for encouraging women to speak up in the classroom, especially in math and science classes. There were some drawbacks, in that it felt a bit like living in a bubble, and didn't prepare us for the harsh realities of sexism in the real world. Because in co-ed classrooms, especially in math and science, it has been proven that teachers call on boys more frequently than girls, therefore discouraging girls from participating.

Many teachers unconsciously perpetuate these gender stereotypes and contribute to the gender-based social environments of middle school and high school. During this time, social approval becomes more and more important for teenagers, and if peers shun math and science, or believe in negative stereotypes about it, then girls will be more likely to abandon their studies in these fields. What we need is to challenge these sexist stereotypes, and to be inclusive of all girls, whether gay or straight. Just as well, in romantic relationships, both teen and adult, and both same-sex and opposite-sex, it is important that partners not feel threatened by the others' intelligence, and are supportive of each other's pursuits.

Insight #86

Telling girls "you're so pretty!" sends the message that their physical appearance is more important than their accomplishments.

THIS GOES BACK to the stereotype of the "smart one" and the "pretty one" being mutually exclusive. As we saw in the last section, peer approval becomes increasingly important in the middle school years. If peers believe that all women in math and science are awkward and dowdy, then girls will be less likely to go into those fields. Because let's face it, being awkward and dowdy is not what most people want in middle school and high school. I should know because I frequently felt awkward and dowdy as a teenager.

These subtle messages often start much earlier, even as early as preschool, with the "princess" culture for little girls. Starting at a young age, a great emphasis is placed upon girls' physical appearance. For example, clothes, hair, makeup, and especially body image. In going back to the TV show *Desperate Housewives*, the character Gabrielle Solis was a fashion model turned housewife and mother of two young girls. She was not very bright but was always concerned with her physical appearance (and her daughters' physical appearances, including their weight). In one episode, she was mentoring young girls for a beauty pageant, and she gave them the advice to avoid math and science because they cause frown lines. I know it was meant in jest, but it was not funny.

There is nothing wrong with taking pride in your physical appearance. In my thirties, I am slowly starting to make more of an effort with clothes, shoes, and makeup than I did when I was in high school, or even in grad school. Nevertheless, it is a sad reality that women are more often judged on their physical appearance than on their merits. Starting at a very young age, women are judged more harshly than men for their physical appearance, including their weight and body image. The cultural emphasis on women's physical appearance contributes to sexual harassment, especially in STEM fields. There was recently an article on Facebook along the theme of "why would women get all dolled up if they didn't want male attention?" There were some excellent points in that article, including the fine line between compliments and harassment.

There is the stereotype in our culture that women in STEM don't care about their physical appearance, and are seen as less feminine because of it. Because of this, the stereotype perpetuates the myth that women in STEM are less feminine, and therefore less desirable. And that is why we place a high premium on females' physical appearance, even as young as three years old. What we need to work on is making these things not mutually exclusive, and to stop judging women by their physical appearance. A woman's physical appearance is not an invitation for harassment or sexual assault, and we need to change that culture, including in STEM fields.

Insight #87

The sheer number of angry tweets in response to ***Doctor Who*** *being a woman is evidence that sexism in STEM is still a problem.*

FOR THOSE OF YOU who are not familiar with *Doctor Who*, it is a classic British science fiction television series that was on the air from 1963-1989, and was since revived in 2005. It involves an extraterrestrial doctor called a "Time Lord," who travels through time in a blue vehicle shaped like a police box, known as the TARDIS. Occasionally, when the season of the show switches, the Doctor switches bodies. For the first 12 seasons, the Doctor has been played by several men, including David Tennant, Matt Smith, and Peter Capaldi.

In summer 2017, it was announced that in a first for the series, the Doctor of the 13th season would be played by a woman, Jodie Whittaker. Shortly after the announcement was made, social media was abuzz with excitement. People took to Facebook, Twitter, and other social media platforms to express their excitement about a history-making first. People wondered why it took so long for the series to cast a female Doctor. Mothers and grandmothers were beyond thrilled that their daughters would get to witness a positive change to the series.

Not everybody reacted positively to the news. Men of all ages took to social media to express their frustration and disappointment, claiming that they would boycott the 13th season of the series. Others claimed that the decision was a publicity stunt, and was designed simply to be "politically correct." Sexist comments began to circulate on social media:

"It's Doctor Who, not Nurse Who!"

"Go to the kitchen and make me a sammich!" (sic)
"I'd bang her!"

In looking at these comments, and their overall sentiments, it's not hard to see the sexism lurking in them, both hostile sexism and benevolent sexism. It is said that art imitates life, and vice versa, and this is such a case. The rampant sexism in response to this casting decision is a reflection of the rampant sexism in STEM fields in real life. Namely, that women are assumed to be less capable in science fields than men, and are expected to take on domestic roles, and are even reduced to sexual objects.

As this book goes to press, the 13th season of *Doctor Who* is about to begin airing. It will be curious to see what plot points and character traits will be ascribed to the character's gender. Several of the tweeters complained that while the Doctor can change bodies, and can even change ages and races, it shouldn't be able to change genders. Luckily, the producers of the show cleared that up by saying that the Time Lord can change with the times. Indeed, the addition of a female Time Lord is a reflection of our changing times.

Insight #88

The legacy of Maryam Mirzakhani is that of a strong female role model in math for generations to come.

IN SUMMER 2017, I received the devastating news of the untimely death of Maryam Mirzakhani, a brilliant female mathematician at Stanford. With her being not much older than me, I was naturally shaken. Upon a deeper reflection, this affected me much more, what with her being not only a woman in mathematics but also a pioneer in her field. In addition to proving several mathematical theorems in her short life and being an excellent professor for several students, she also represented a lot more. Namely, she represented the struggles of being one of very few women in a male-dominated field, and still succeeding in spite of the odds against her.

She was born in Iran in 1977 and grew up in an Iranian household, where not many women were involved in math and science. Needless to say, she faced a lot of sexism growing up, and yet still triumphed in earning her Ph.D. in mathematics. Remember the concept of intersectionality when it comes to underprivileged groups? She not only had to face sexism but also Islamophobia.

At the age of 37, in 2014, she earned the Fields Medal, the most prestigious award in mathematics. And just like it took over 40 years for *Doctor Who* to cast a female Time Lord, it took over 80 years for this award to be presented to a woman.

This begs the question: why did it take so long for a woman to win this award? It certainly wasn't because men are inherently better at math or more interested in math. It was because of a lack of opportunities for women in math, and social and political pressures against women.

Although Mirzakhani lived a short life, she paves the way for many women mathematicians in the future. Not only women mathematicians, but also Muslim women mathematicians and foreign-born mathematicians. Because of her influence, many girls will know that it is possible for them to go into mathematics, and to make lasting contributions to the field, in spite of sexist challenges. Mirzakhani left behind a young daughter, who described her mother's work in mathematics as "painting." Mirzakhani's legacy shows the value of taking your time in mathematics, saying that "you have to spend some energy and effort to see the beauty of math."

Because of her legacy, the beauty of math will be accessible to many young women worldwide. Although she has so far been the only woman to win such a prestigious award in mathematics, it is unlikely that she will be the only one. While she might have lived a short life, her legacy will continue on in the future generations of women who pursue the field of mathematics. Because of that, Mirzakhani was a pioneer in the true sense of the word.

IT'S NOT TOO LATE

(Insights 89-99)

Insight #89

"I wish I would have..." statements give your present power away to the past.

IF I HAD A DOLLAR for every time an adult (usually a woman) said to me, "I wish you would have been around when I was young!" I would be able to retire in Paris before my 40th birthday.

There was once an episode of *Sex and the City* titled "Coulda, Woulda, Shoulda." While it is unlikely that Carrie Bradshaw studied advanced math, this title certainly applies here. So many adults have regrets about the past. If only this, if only that. Yet, we can't change the past. According to Oprah, the definition of forgiveness is giving up hope that the past could be any different. On the upside, the present and future can certainly be different from the past, but if we don't come to terms with the past, then it will affect the present and future, most likely in a negative way. Whenever I hear that statement, I think to myself, "Your childhood (or high school days or college days) might not have been perfect, but it's over. No use dwelling on the past." And yet, "you wouldn't be telling me this story if it wasn't still affecting your life in the present."

There is a difference between processing the past and rehashing the past. We want to encourage the former, but become aware of and reframe the latter. What is the difference? We can't change the facts or events of the past, but we can change our thoughts and feelings about them. In a sense, rehashing the past means going over the events of the past with the same old thoughts and feelings, but reframing the past means going over the events of the past with new thoughts and feelings. Rehashing the past keeps you stuck in the past, whereas reframing the past helps you to move forward.

When an adult says that they wish I would have been around when they were taking math classes in high school, I occasionally respond by asking, "hypothetically, how would things be different today if I had been around when you were in high school?" This brings them back to the present and helps them to reframe the past. Thus, it helps them to realize that while they can't change the past, their old story about the past is still affecting the present, and is holding them back. How does it affect the present? We will be going into this in much more detail later, but let me leave you with the three F's of why adult math shame matters: finances, family, and forsaken dreams. All three of these affect the present and yet are rooted in the past. If we don't make peace with the past, then it is going to negatively affect not only the present, but also the future.

Remember, math shame is an intergenerational wound that requires intergenerational healing. That is, if a parent has math shame, then there is a good chance that their child will develop math shame as well. If you don't heal your past for the sake of yourself, perhaps consider doing so for the sake of your children.

Insight #90

Your relationship with math does not end when you stop being a student.

IT IS UNFORTUNATE that most adults think that my work is no longer relevant to them. This is especially true for adults who are parents and/or teachers. Because the fact of the matter is, my work is very relevant to adults, and not just parents and teachers. That is because my work explores your relationship with math. I have a friend who works as a holistic nutrition coach (as opposed to a weight loss coach), and she focuses her work on healing her clients' relationships with food, past and present. It is the same principle as my work.

What is your relationship with math? Most of you probably have a love-hate relationship with math. Or to be perfectly honest, a hate-hate relationship with math. You would much rather leave math in the past, where you think it belongs. How is that affecting you today? How is it affecting your career? How is it affecting your finances? If you are a parent or teacher, how is it affecting your children and/or students?

When it comes to adults' relationships with math, there are three areas that I like to cover, which I call the three F's: finances, family, and forsaken dreams. Finances are pretty self-explanatory. Everyone has a relationship with money. I often like to say that a negative relationship with math in childhood can manifest as a negative relationship with money in adulthood. Although we as women desire more money, there are always some subconscious blocks. Sometimes, those subconscious blocks have to do with the stories we tell ourselves from childhood. For example, "I'm not a math person," or "I was never good with numbers." This then translates into an anxiety about managing money, because people think that managing money means working with numbers. Well, technically, there are numbers involved, but it's nothing like high school algebra.

Speaking of high school algebra (or even elementary school addition), if you have kids or teenagers (or even young adults), your attitudes about math could be affecting their attitudes about math. Thus, if you have a negative relationship with math, then they are likely to have a negative relationship with math as well. The first step to healing that in them is to heal your own relationship with math.

Insight #91

Your past in math does not dictate your future in math.

IN OUR EDUCATIONAL system of grades, test scores, competition, and ranking, it is a common belief that the best predictor of future math performance is past math performance. In other words, to predict how a child will perform in math in middle school, look at elementary school performance. To predict how a teenager will perform in math in high school, look at middle school. To predict how a young adult will perform in math in college, look at high school.

While this may be true in most cases, it is not true in all cases. What's more, this model does not give students room for growth and change. It is a common saying in statistics that correlation does not imply causation. In other words, just because a student has poor grades in middle school math does not mean that they will get poor grades in high school math, and vice versa. Students can and do change, and can sometimes make extraordinary leaps and bounds within the course of less than a year.

Take past grades and test scores with a grain of salt. Grades and test scores are not always an accurate reflection of performance or achievement in math. Just as well, this train of thought implies the fixed mindset and puts students in boxes. Remember the part about labeling a child as an "A student," a "B student," a "C student," and so forth? Same principle here. Just because a student had a rocky past in math does not mean that they can't have a bright future.

The same principle can apply to adults who haven't seen the inside of a classroom in years (or decades!). Just because they had bad experiences in math as a child does not mean that they can't have good experiences in math as an adult. Unfortunately, we often allow our experiences in the past to cloud our experiences in the present. The same can apply to parents and teachers. For example, parents and teachers might compare a younger sibling to an older sibling. Or they might look at previous grades in math, and assume that a child will perform poorly in their class. It is these assumptions that can cloud their experiences and can affect the experiences of the child in question.

No matter what your age, you deserve a fresh start in math. Things can change in an instant if you let them. In one of my online groups, we have a hashtag, #UpUntilNow. That is, with this hashtag, we are actively aware of changing things, and of making a fresh start for the future. Let's apply this hashtag to math. For example, #UpUntilNow, I believed that I was bad at math. Or #UpUntilNow, I thought that I wasn't a math person. Notice how empowering that feels? So why don't you try that with math? Make a list of the things that are going to be different in the future.

Insight #92

You cannot change your past. But you can change how your past affects you today and in the future.

ACCORDING TO OPRAH WINFREY, the definition of forgiveness is giving up hope that the past could be any different. For both teenagers and adults, I'm sure that there are a lot of experiences in math that you need to forgive. Not only forgiving those who have affected you in the past (i.e., parents, teachers, etc.) but more importantly, forgiving yourself. Once again, I see a lot of hurt from the past being carried into the present and the future. Adults often wish that their childhood would have been different.

Let me ask you a hypothetical question: Hypothetically, if I was around when you were a teenager, how would things be different today? We can apply this concept to virtually anything, and not just math. Hypothetically, if the past had been different, how would the present and the future be any different? While it's fun to dream, there's a point to this exercise. That hypothetical present or future that you just described is very possible, in spite of your past. Not only in spite of your past, but possibly because of your past. Namely, the wisdom you have gained from your past. We all have a past, and we all are powerless to change our past, including our past in math. We have power in the present. We can choose how our past affects our present and our future.

Now that you have dreamed up your hypothetical present and future, let's clear the past as an obstacle. What limiting beliefs are keeping your past as an obstacle? In going back to the definition of forgiveness, we need to release the painful experiences in the past, and we also need to forgive ourselves. Remember, we did the best we could with what we had. Even if you're as young as eleven years old at the time you're reading this, you can still practice forgiveness of your past. Just remember that you didn't know then what you know now and that what happened then can't be any different.

Let's have some fun: how can we make your hypothetical future and present happen? What needs to happen in order for those dreams to come true? Here, we are in a state of possibility, now that we have removed the past as an obstacle. In going back to telling your past stories to your teenagers, it would be well worth it for you to do some forgiveness surrounding these stories before you tell them to your children or students. Because not forgiving someone is like drinking poison and expecting the other person to die. Don't forgive someone because they were right and you were wrong; forgive someone for your peace of mind. This also goes for all the people who may have contributed to your negative feelings about math in the past.

Insight #93

Math shame is not only a school issue, but also a life issue.

IT IS SAD THAT many people have a very limited view of the scope of my services. It is commonly thought of as focusing exclusively on school-age children, and focusing exclusively on the math material that is taught in school. However, my work addresses a much deeper societal issue that goes beyond the classroom and affects all ages.

Yes, I do work with students who are currently in school, and I do give speeches to high schools and colleges. But did it ever occur to you what happens when these students graduate? For many of them, they abandon their math studies as soon as they graduate (or complete the minimum requirements), and never look back, except to regret lost opportunities. Has it ever occurred to you that their memories of math shame might be affecting them for years after they leave the classroom?

Math goes beyond the classroom. By extension, math shame also goes beyond the classroom. Think about it this way: if anyone over the age of 8 (maybe 10 in some cultures) admitted that they didn't know how to read, people would judge them as illiterate and would suggest that they get some reading lessons. For math, it has become somewhat of a female bonding ritual to say that you're bad at math. This extends to all ages of women, going far beyond the classroom.

If you think about it, according to the "lock-step" model of society, school is only 18-22 years of a person's life, or perhaps up to 26 if they opt for graduate or professional school. There is a great deal of life after this age range, and math does not simply disappear when one leaves the classroom.

While the person might have abandoned formal math education, the math shame still stays with them in many cases, and unconsciously affects them. Just like with reading and writing, math has many applications to everyday life as an adult, not the least of which is one's financial life.

Even when calculating a bill at a restaurant or bar, math shame often shows up. Many people panic when the bill arrives, and whip out their iPhones. I once saw a meme on Facebook about teachers in the '70s and '80s saying, "you won't have a calculator with you at all times," and kids of the new millennium saying, "we told you so!" (In reference to iPhones). But it's sad that so many people become reliant on calculators as a crutch, and worse, feel ashamed about it.

Math shame promotes the fixed mindset. Just as a refresher, the fixed mindset believes in abilities that are set in stone, and that cannot be changed. A perfect example of this is the "it's too late" mentality that many adults have when it comes to improving their math skills.

Insight # 94

Behind every woman who says, "I wish I would have taken more math when I was young" is a story of loss and regret.

AT THE RISK OF sounding like a broken record, let's revisit the "I wish you would have been around when I was young!" phenomenon.

The youngest person who said this to me was a whopping 19 years old at the time. Although she seems happy and well-adjusted as a beauty student and a secretary and assistant for a senior stylist, my heart still aches for her. Just two years before, she was suffering in her math classes in high school, and couldn't wait to squeak by with the minimum math requirements so that she could finally graduate from high school and start beauty school. An older woman, who is celebrating her 90th birthday as this book goes to press, says the same thing.

In spite of their vast differences in age and demographics, these two women both illustrate the sadness of loss and regret when it comes to math. The sad reality is that many of these women simply believe that it is too late for them. Not only too late for them to take more math classes and re-learn the math that they were missing in school, but also too late for them to release the shame that they have been carrying around for years (or even decades!). Most of those women do not want to revisit their painful memories surrounding math, because they have it drilled into their heads that math is evil, and that math is a painful experience. However, it is those painful experiences that they need to heal from. Even if these women don't take more math classes, this healing can be a transformative experience. Not only for them, but also for their daughters and granddaughters, and for women as a whole.

Let's revisit the "it's too late" limiting belief. Well, for starters, there was a woman who earned her first undergraduate degree at 96, and another woman who got her Ph.D. at 107, so if they can do it, anyone can. After all, according to the Mark Twain quote, we more often regret the things we don't do, rather than the things we do.

A good resource on challenging the "it's too late" limiting belief is the book *The Artist's Way* by Julia Cameron. One of my favorite Julia Cameron quotes begins with the objection, "do you know how old I'll be by the time I learn to play the piano?" Julia's response? "The same age you'll be if you don't." Same principle with math, which includes not only learning the math material, but also letting go of the math shame.

Insight #95

Being vulnerable about math anxiety and math shame can be difficult, but ultimately rewarding, for both you and your children or students.

ONCE AGAIN, I KNOW that vulnerability and math are usually seen as a contradiction. In other words, most people think that they don't go together. After all, math is a very logical, head-centered subject, and vulnerability has to do with the heart. Many people think that exposing one's vulnerability means opening oneself up to being seen as weak and immature. Crying is frequently seen as a sign of weakness for anyone over the age of twelve, whether male or female (although the crying is usually treated differently based on gender).

Crying, and being vulnerable in general, can be very healing, and can invite others onto your healing journey. I know we've talked about Brene Brown before, but she is a great proponent of vulnerability. One of her major themes is that shame tends to live in secrecy. She argues that vulnerability is a tool not only of healing but also of connection. One of my mentors always used to say, "just because you show your vulnerability does not mean you lose your credibility." Once again, this might seem like a contradiction, because most people think that it is a sign of weakness for an adult or a teenager to show any sign of emotion.

While vulnerability can be difficult, it can also be rewarding in terms of healing your math anxiety and math shame. Not only that, but it can also be a great opportunity to connect with your students, children, and peers. When one is not allowed to be vulnerable, it negatively affects their mindset and mood. After all, emotions tend to get bottled up, and the more we try to bottle them up, the more they snowball, and the more they affect us. What's more, there's a define sexist divide when it comes to expressing one's emotions, with teenage boys and adult men, gay and straight alike, being especially discouraged from being vulnerable.

Another difference between me and most math tutors is that I encourage vulnerability. For example, if a student starts crying in front of a typical tutor, the tutor would likely say, "Are you OK? Don't cry!" and hand the student a tissue. I, on the other hand, believe in the healing power of tears and will allow the student the emotional release. After all, math can trigger many painful emotions. Similar to crying, most parents and teachers also discourage teenagers from using foul language. Once again, I'm the opposite. Like crying, swearing can be a powerful emotional release. The key is to express both in the appropriate context. If your teenager wants to shed some tears and shout some F-bombs about a math test, you've come to the right place! After all, both crying and swearing can work wonders in shifting emotional energy.

Insight # 96

Staying present in the present instead of dwelling on the past is critical to ending math shame. In order to release your past, you must first face it.

MANY TIMES, SHAME triggers can bring up painful memories of the past. They can also trigger anxiety about the future and fear of the unknown. For example, "will I get into college?" "Will I get a good job?" "Will I be stuck working at Starbucks for the rest of my life?" Then again, an adult might lament, "I wish I would have done this differently." It has been said that if you're living in the past, you're depressed, and if you're living in the future, you're anxious. If and only if you're living in the present, then you're at peace. Not very many students, parents, or teachers are living at peace when it comes to math.

In going back to my parallel struggles with weight, I have often felt like I was in the middle of a war with food and my body. It was not until I was introduced to the concept of intuitive eating that I began to make peace with food. Likewise, tutoring often sets up a "war" with math, just like dieting sets up a "war" with food. Thus, working through the material without addressing the shame only adds fuel to the fire of the war. Many times, when we're fighting the war, our mind wanders to the past or the future.

We can't change the past. What's done is done. Similarly, we can't control the future. We can only control ourselves and our own reactions. In order to stay present in the present, we must first face our past, as painful as it may be. On the other hand, facing your past does not mean dwelling on it.

There are many ways to stay focused in the present, one of which is meditation. Contrary to popular belief, mediation does not have to be religious, and it does not have to be sitting and chanting. It is simply a form of gathering your thoughts, and of not letting your thoughts control you. It is a form of awareness of one's own thoughts. Similarly, journaling and spending time in nature can have similar effects.

In applying this to math, it is vital to stay present in the present, and just focus on the problem, assignment, or test that is at hand. Many times, our minds have a tendency to wander. For example, if a problem becomes frustrating, our minds might wander to a similarly frustrating experience in the past, and it might trigger similar feelings of shame and doubt. Or your mind might wander to the future, and your anxiety about a test, or about college admissions. The key here is to become aware of these thoughts and let them go. If you get a thought that is particularly triggering, especially about a painful past experience, it might be worth it to do some extra journaling, healing, and release work around it.

Insight #97

Math shame can affect people for a lifetime—throughout childhood, adolescence, and adulthood.

I KNOW I'M SOUNDING like a broken record, but let's go back to the old "I wish you would have been around when I was young!" phenomenon. Most people who say this have unhealed math shame that they carried into adulthood. Most of these cases of math shame started during childhood. Upon further discussion with these women, we can usually find a beginning to this math shame in childhood, and often before the age of seven.

As an example, one woman told the story of when she was six years old, and she was selling Girl Scout cookies to a lady at church. She felt put on the spot when the lady asked her to count her change, and she froze up. That experience stayed with her for years to come, and she never truly healed it. This woman is not alone.

So many cases of math shame start with childhood experiences. When most people hear the term "math shame," they think of childhood and school, and they assume that as adults, it's too late for them to heal. The truth is, math shame is a life issue, and not a school issue. That being said, it almost always begins in childhood, and more often than not, begins with school. I am often asked about Common Core. The truth is, even before Common Core, the school system was a shame-based place, even as early as the 1950s. Adults of all ages can have shame-based memories of their days in school, and this shame can last a lifetime.

What we really need to do is to change the shame-based culture of the education system. Unfortunately, we cannot do that without a major overhaul. What we can do is to change our response to this shame-based system. In order to start, we need to recognize that math shame is an issue for all ages, and can last a lifetime. So literally, childhood experiences can shape our perceptions for years to come.

Fortunately, our childhood does not have to dictate the rest of our lives. This is true not just in math, but in shame in other areas. In going back to my parallel experiences with shame in social situations and weight and body image, I have a lot of childhood experiences that I still have not healed. I now know that it is a journey to heal my childhood shame and that in my thirties, it is not too late for me to reframe these childhood experiences. By extension, it is not too late for you to heal your childhood math shame and to make a positive difference in your adulthood.

Insight #98

It is never too late for anyone to overcome math shame.

"I WISH YOU WOULD have been around when I was young!" Sound familiar? Every time I hear those words (or something to that effect), my heart breaks a little. Not only because it represents a lost opportunity, but because the person saying the statement most likely thinks that it is too late for her (or in some cases, him) to heal the emotions underlying this statement and to move forward from it.

Some of you might have had career dreams that you gave up because they involved more math than you were willing to handle at the time. For example, a doctor, nurse, psychologist, engineer, architect, or computer programmer. You might be wondering "what if?" Then you remind yourself that you have a family, a job, a house, and other responsibilities, so you toss your old dream aside and lament how you wish the past would have been different. Remember what Oprah said about forgiveness? Until you make peace with the past, it will continue to chase after you.

The belief that it's too late is a limiting belief for a lot of women (and men) for a lot of things and not just math. This can also apply to younger people as well. In our society, we have a prescribed timeline of sorts, wherein if you don't achieve certain milestones by a certain arbitrary age, then you're labeled as a "failure" or "behind," which is absolutely not true. Unfortunately, many people are still bound by this belief. After all, the youngest person who said the most common statement that I hear was only 19 years old at the time.

Even years after these incidents happen, the shame can linger. And it can affect not just our work with numbers in the present (i.e., calculating taxes or a bill), but can also affect other areas of our lives. If you think about it, math shame is a self-esteem issue. And self-esteem affects all areas of a person's life, including finances and family. And remember, parents set the tone for children, who are the future potential STEM workforce. A great summary of my work is that the parents' past affects the children's present, and the children's present affects the world's future. In other words, as a parent, you can have an impact not only on your family, but also on the world.

I promise you, it is never too late. Think about what you could gain from letting go of your math shame. And along the "it's not too late" theme, I'm reminded of Julia Cameron's book *The Artist's Way*. She recently published a new edition, titled *It's Never Too Late to Begin Again*, aimed at adults who are middle-aged or older, usually in retirement age. Indeed, many adults who come to me are "empty nesters" who chase after dreams that they abandoned in high school or college.

I promise, it is never too late for you.

Insight #99

It's okay if math is not your favorite subject.

LET'S END OUR 99 Insights by saying that it's okay if math is not your favorite subject. We all have things that we love and hate, and those things are different for different people. Diversity is the beauty of the world. For example, I love coffee, anything purple, Disney movies, anything in Oregon and Washington, and obviously math. I have a friend who loves sweet tea, the color green, horror movies, the South, and prefers reading and English. We're still friends because we can agree to disagree. Just because something's not your favorite does not mean that you can't like or appreciate it.

Same principle with math. Remember, there's no such thing as a "math person." Math is a skill that can be grown and developed, and takes practice. You don't have to love it in order to get through it. For example, there are many tasks that I don't enjoy, but that I feel better once I get them done. Even just 15 minutes can go a long way.

Math might not be one of your ultimate passions, but it might have a purpose in leading to your passions. It could be something as simple as graduating from high school. Obviously, most high schools require math in order to graduate, so that could be a motivating factor. You don't have to love math, but you can certainly change your perspective on it. In other words, it is a means to an end, and if you keep your eye on the prize, it becomes less painful, even if it's not your favorite.

In other words, you don't have to be passionate about math, but a small attitude shift can go a long way. This is also true for parents, who might or might not have seen the math material for years (or decades!). I have witnessed many inspiring stories of grown women who went back to school to study math and had an entirely different experience the second time around. On the other hand, you might surprise yourself. You might find math to be enjoyable. I know this seems unfathomable to many of you, but keep an open mind. Some people who hated algebra loved geometry, and vice versa. Sometimes, people who hated math in high school loved it in college. All I'm saying is to keep an open mind, but don't force it. You don't have to force yourself to feel passionate about something that you would rather not do. Then again, you might surprise yourself.

With all being said and done, after you've done the inner work on healing your math anxiety and math shame, you don't have to suddenly love the subject. But math is a part of our lives and is a stepping stone to many great career opportunities for the 2020s and beyond. So once again, keep an open mind. You just might surprise yourself.

WRAP-UP

Parting Words

WE HAVE BEEN ON quite a journey together, through 99 Insights. Although it may not be what you expected, I hope you got something out of this. Your something might not look like your best friend's something or your child's something or your colleague's something, but I hope there was something for everyone.

So, what's next? I know this is a lot to digest, and a lot consider. After all, most of these principles go against the common beliefs in the educational system today. My advice is to take what resonates and leave the rest. It's okay to be uncertain about the future. After all, one of my spiritual mentors says to "dance in the mystery." In other words, to be okay with not knowing the outcome, and to be okay with whatever happens. Trust that it will all work out, and have faith in the Universe.

That's what this is all about. Having a faith-based approach to math education, instead of the common fear-based approach. This approach is not only fear-based but also shame-based. Unfortunately, we still live in a society that promotes fear-based and shame-based messages about math. What to do? Think of the 99 Insights as a toolbox. Use the tools that speak to you. Even the tools that might or might not speak to you today might speak to you six months from now, one year from now, or even five years from now. It is a journey, not a destination.

Use this as a dialogue with your kids/teens or students. Be open with your feelings about math, and invite them to be open with their feelings about math. Remember, there are no right or wrong feelings about math. Part of the beauty is to honor all feelings and to trust the process. While this can be challenging at first, the rewards can be immense. All in all, trust the process and trust in oneself.

This is not the end of your journey. This is just the beginning. This can apply not only to kids and teens but also to adults. After all, they call a graduation ceremony a "commencement" for a reason. That is because it is not only an ending but also a beginning. Remember, all endings are beginnings.

I invite you to see this as an opportunity to create a new beginning in your relationship to math, no matter what your age.

About the Author

AS A MATH EDUCATION professional since 2006, Molli the Math Lady has developed the revolutionary, holistic 7M approach to banishing math anxiety and its deeper cousin, math shame. She received her master's degree in human development from Pacific Oaks College, and wrote her master's thesis on the topic of gender and math anxiety. She has since incorporated spiritual training into her work, and is an alum of Gabrielle Bernstein's Spirit Junkie Masterclass. She lives in Washington, but is always willing to travel to speak on these topics and more.

Connect with Molli

WEBSITE: MollitheMathLady.com

INSTAGRAM: @MollitheMathLady

FACEBOOK: @MollitheMathLady

Molli also speaks on the following topics:

- Math Anxiety (middle school, high school, and college)
- Math Shame, and how it affects adults
- The importance of STEM (science, technology, engineering, and math) careers
- Women in STEM
- STEM and the arts
- Mindfulness and meditation for math anxiety
- Spiritual healing of math shame
- The importance of parents and teachers in conquering math anxiety
- Social and political aspects of STEM

For speaking requests, please visit MollitheMathLady.com

Bonus: Sneak Preview

Beyond Math Anxiety 2: 99 More Insights (and a Calculation's Still Not One!)

Releasing in 2019

You might or might not be familiar with the term "gifted", or the term "highly sensitive person." There are a lot of myths and misconceptions about both. You might see a gifted student as a bookworm and a recluse who is pressured by a pushy parent, like the British grandmother in the 2017 film *Gifted*. Or you might see a highly sensitive person as an overly irrational, emotional person who needs to "toughen up" and develop a "thick skin." Both descriptions are untrue, and these myths will be addressed in my new book, releasing in 2019. These myths harm those students (and parents!) who truly are gifted and/or highly sensitive, and can contribute to their challenges with math anxiety and math shame.

In addition, we will discuss learning challenges such as ADHD, dyslexia, and autism/Asperger's, and how these can affect math anxiety and math shame. Just as well, we will address mental health issues such as depression and anxiety disorders. These populations can have extra challenges with math anxiety and math shame, because they often feel these emotions more intensely, and can also be more sensitive to others' feelings. As a result, it becomes a double-edged sword of sorts, with them struggling to manage their own feelings about math, as well as dealing with their reactions to others' feelings. In addition, we will address the common misconception that gifted students are always high achievers. Paradoxically, gifted students often need more support in math, not less. This is true whether they are "ahead" or "behind" according to traditional grade-based standards. In addition, the unique emotional needs of the gifted and creative will be discussed. Other issues such as Dabrowski's over-excitabilities and asynchronous development will also be discussed.

To top this off, highly sensitive and gifted children and teenagers often (but not always) come from highly sensitive and gifted parents. Thus, the parents will very likely be navigating their own journey with these unique needs along with their children. In addition, burnout is also a common issue for highly sensitive people, and so boundaries and self-care will be explored at length. All 99 of the original insights can be applied to these populations. Sometimes these populations need a little extra support and understanding. This is where the 99 more insights come into play.

Even if you don't think you identify with these populations, you probably know someone who does. All in all, the more sensitivity and understanding we have, the better. Just as well, these populations are essential to the pipeline of the future of STEM, and without these special minds, we will lose a lot. Unfortunately, our system is not very understanding of the needs of these populations, so the 99 new insights, in addition to the 99 original insights, will shed some light.

References

- Baron-Cohen, Simon. *The Essential Difference: Male and Female Brains and the Truth About Autism*. Basic Books, 2003. New York, NY.
- Beilock, Sian L. "Female Teachers' Math Anxiety Affects Girls' Math Achievement." *Proceedings of the National Academy of Sciences USA*. February 2, 2010.
- Bergland, Christopher. "Why Is Physical Activity So Good for Your Brain?" www.psychologytoday.com/us/blog/the-athletes-way/201409/why-is-physical-activity-so-good-your-brain.
- Bernstein, Gabrielle. *The Universe Has Your Back: How to Feel Safe and Trust Your Life No Matter What*. Hay House, 2016. London.
- Boaler, Jo. *Mathematical Mindsets: Unleashing Students' Potential through Creative Math, Inspiring Messages, and Innovative Teaching*. Jossey-Bass & Pfeiffer Imprints, 2016. San Francisco, CA.
- "Brain Halves Interact Differently with Each Other." www.blog.pnas.org/2013/08/brain-halves-interact-differently-with-each-other/.
- Brown, Brene. *Daring Greatly: How the Courage to be Vulnerable Transforms the Way We Live, Love, Parent, and Lead*. Penguin Life, 2015. London.
- Burgess, Heidi. "I-Messages and You-Messages." February 28, 2017. www.beyondintractability.org/essay/i-messages.
- Cameron, Julia. *The Artist's Way: A Spiritual Path to Higher Creativity*.: Tarcher Peregree, 1992. New York, NY.
- Cameron, Julia and Emma Lively. *It's Never Too Late to Begin Again: Discovering Creativity and Meaning at Midlife and Beyond*. Tarcher Perigee, 2016. New York, NY.
- Canfield, Jack and Janet Switzer. *The Success Principles: How to Get from Where You Are to Where You Want to Be*. William Morrow, 2015. New York, NY.
- Clinton, Hillary Rodham. *What Happened*. Simon & Schuster, 2017. New York, NY.

- Dweck, Carol S. *Mindset: The New Psychology of Success*. Ballantine Books, 2008. New York, NY.
- Foran, C. "Donald Trump and the Triumph of Climate-Change Denial." December 25, 2016. www.theatlantic.com/politics/archive/2016/12/donald-trump-climate-change-skeptic-denial/510359/.
- "Fractal Patterns in Nature and Art Are Aesthetically Pleasing and Stress-Reducing." March 31, 2017. www.smithsonianmag.com/innovation/fractal-patterns-nature-and-art-are-aesthetically-pleasing-and-stress-reducing-180962738/.
- Gilligan, Carol. *In a Different Voice: Psychological Theory and Women's Development*. Harvard University Press, 1993. Cambridge, MA.
- Grandin, Temple. *Thinking in Pictures: And Other Reports from My Life with Autism*. Vintage Books, 2006. New York, NY.
- Hatcher-Skeers, Mary. "Sorry, Men, The Gender Equity Problem in Science is Not Solved." *Scripps College Magazine*, 2008. http://magazine.scrippscollege.edu/2008-summer/sorry-men-the-gender-equity-problem-in-science-is-not-solved.
- Hoffman, Jan. "Math Anxiety? A Reporter Knows the Subject All Too Well." August 25, 2015. www.nytimes.com/2015/08/25/insider/math-anxiety-a-reporter-knows-the-subject-all-too-well.html.
- Iabichela, M. "You Only Get What You Have the Capacity to Receive." www.infinitereceiving.com/.
- "Introducing Vibration: From Science to Energy Healing." www.blog.mindvalley.com/vibration/.
- Kerr, Barbara. *Smart Girls: A New Psychology of Girls, Women, and Giftedness*. Great Potential Press, Inc, 1997. Scottsdale, AZ.
- LaPorte, Danielle. *The Desire Map: A Guide to Creating Goals with Soul*. Sounds True, 2014. Boulder, CO.
- Lawson Davis, Joy. *Bright, Talented, and Black: A Guide for Families of African American Gifted Learners*. Great Potential Press, Inc., 2010. Scottsdale, AZ.
- "Limbic System." www.sciencedirect.com/topics/agricultural-and-biological-sciences/limbic-system.

- Lyons, I. M. and Sian. L. Beilock. "When Math Hurts: Math Anxiety Predicts Pain Network Activation in Anticipation of Doing Math." www.journals.plos.org/plosone/article?id=10.1371/journal.pone.0048076.
- "Mathematics Standards." www.corestandards.org/Math/.
- Mazur, Joseph. "How Pi Proved It Deserved a National Holiday." March 14, 2016. www.slate.com/articles/health_and_science/science/2014/03/pi_day_history_perfect_symmetry_a_mathematical_constant_wave_formulas_and.html.
- McKellar, Danica. *Math Doesn't Suck: How to Survive Middle School Math Without Losing Your Mind or Breaking a Nail.* The Penguin Group, 2008. New York, NY.
- McKellar, Danica. *Kiss My Math: Showing Pre-Algebra Who's Boss.* The Penguin Group, 2009. New York, NY.
- McKellar, Danica. *Hot X: Algebra Exposed!* The Penguin Group, 2011. New York, NY.
- McKellar, Danica. *Girls Get Curves: Geometry Takes Shape.* The Penguin Group, 2012. New York, NY.
- McKellar, Danica. "Danica McKellar's 5 Tips to Conquer Math Phobia, for Your Kids' Sake." *MSNBC.* 2012. www.moms.today.msnbc.msn.com/_news/2012/08/13/13221807-danica-mckellars-5-tips-to-conquer-math-phobia-for-your-kids-sake?lite.
- National Science Foundation. "Recommendation Letters May Be Costing Women Jobs, Promotions." 2010. www.usnews.com/science/articles/2010/11/12/recommendation-letters-may-be-costing-women-jobs-promotions,
- Obama, Barack. *The State of the Union Address.* 2012.
- Petroff, Alanna. "The Exact Age When Girls Lose Interest in Science and Math." www.money.cnn.com/2017/02/28/technology/girls-math-science-engineering/index.html.
- Pipher, Mary. *Reviving Ophelia.* Penguin Group, Inc, 1994. New York, NY.
- Richardson, F. and R. Suinn. "The Mathematics Anxiety Rating Scale: Psychometric Data." *Journal of Counseling Psychology,* 1972.

- Rotenstein, Lisa S., Marco A. Ramos, and Matthew Torre, MD. "Prevalence of Depression and Suicidal Ideation Among Medical Students." December 06, 2016. www.jamanetwork.com/journals/jama/article-abstract/2589340.
- Ruf, D. L. *5 Levels of Gifted: School Issues and Educational Options.* Great Potential Press, 2009. Scottsdale, AZ.
- Russell, Cristine. Confronting Sexual Harassment in Science. October 27, 2017.www.scientificamerican.com/article/confronting-sexual-harassment-in-science/.
- Santrock, J. *Adolescence (13th Ed.).* McGraw-Hill Companies, 2010. New York, NY.
- "Scientific Consensus: Earth's Climate is Warming." February 08, 2018. Retrieved from https://climate.nasa.gov/scientific-consensus/.
- Seaman, R. "Seattle Life Coach Training." www.seattlelifecoachtraining.com/.
- Silverman, L. K. "The Many Faces of Perfectionism." www.giftedhome-schoolers.org/resources/parent-and-professional-resources/articles/issues-in-gifted-education/the-many-faces-of-perfectionism/.
- Simon, M. Unpublished Doctoral Dissertation. 1989.
- "STEM to STEAM." www.stemtosteam.org/.
- Stevens, H. "Why Do Women Get All Attractive if They Don't Want to be Harassed? Glad You Asked." December 20, 2017. www.baltimoresun.com/news/ct-life-stevens-sunday-why-do-women-make-themselves-attractive-1105-story.html.
- Suri, Manil. "Why Is Science So Straight?" September 05, 2015. www.nytimes.com/2015/09/05/opinion/manil-suri-why-is-science-so-straight.html.
- Tannenbaum, Melanie. "The Problem When Sexism Just Sounds So Darn Friendly...." April 02, 2013. www.blogs.scientificamerican.com/psysociety/benevolent-sexism/.
- Tobias, S. *Overcoming Math Anxiety.* W.W. Norton & Company, Inc., 1978. New York, NY.
- Tobias, S. *Overcoming Math Anxiety.* W.W. Norton & Company, Inc., 1993. New York, NY.

- Vangelova, Luba. "5-Year-Olds Can Learn Calculus." March 03, 2014. www.theatlantic.com/education/archive/2014/03/5-year-olds-can-learn-calculus/284124/.
- Vierling-Claassen, Angela. "From Math Shame to Math Liberation." December 14, 2012. www.angelavc.wordpress.com/2012/12/14/math-shame-math-liberation/.
- Wheeling, K., J. Cohen, R. Stone, G. Vogel, L. T. Baron, and P. Voosen. "The Brains of Men and Women Aren't Really that Different, Study Finds." December 09, 2017. www.sciencemag.org/news/2015/11/brains-men-and-women-aren-t-really-different-study-finds.
- Wigfield, A. "Math Anxiety in Elementary and Secondary School Students." *Journal of Educational Psychology*. 1988. Vol. 80, No. 2, pp. 210-216.
- Williams, H. R. "A Winning Parental Leave Policy Can Be Surprisingly Simple." February 27, 2018. www.hbr.org/2017/07/a-winning-parental-leave-policy-can-be-surprisingly-simple.

CPSIA information can be obtained
at www.ICGtesting.com
Printed in the USA
BVHW03s1819270918
528693BV00001B/22/P